Flying Saucer Technology

Bill Rose

First published 2011

ISBN 978 1 857803 31 0

Published by Ian Allan Publishing

an imprint of Ian Allan Publishing Ltd,
Hersham, Surrey KT12 4RG.

Printed in England by Ian Allan Printing Ltd,
Hersham, Surrey KT12 4RG.

Visit the Ian Allan Publishing website at **www.ianallanpublishing.com**
Distributed in the United States of America and Canada by BookMasters Distribution Services.

Photograph on title page:**A simulated image of two
unidentified bright lights in the night sky, typical of
numerous UFO sightings.** *Bill Rose*

FLYING SAUCER
TECHNOLOGY

BILL ROSE

MIDLAND

Contents

Preface

'These airplanes we have today are no more than a perfection of a toy made of paper children use to play with. My opinion is we should search for a completely different flying machine, based on other flying principles. I consider the aircraft of the future that which will take off vertically, fly as usual and land vertically. This flying machine should have no parts in movement. The idea came from the huge power of the cyclones.'

Henri Coanda, Aerodynamicist (1967)

Although I had occasionally seen illustrations of flying saucers in comic books, my formal introduction to these craft took place during the late 1950s, when my father took me to see a science fiction (SF) movie titled *Forbidden Planet* (released in 1956). Set several hundred years in the future, one of the central features of this classic movie was the very advanced 'United Planets' C-57D disc-shaped interstellar spacecraft.

I was immediately won over by the elegance of this impressive vehicle. However, the idea of it being man-made was rather novel. Up to that point, flying saucers portrayed in SF usually carried aliens to the Earth, with writers and film producers capitalising on widespread public interest in unidentified flying objects (UFOs). Although there are accounts of (what might be considered) UFO sightings dating back to biblical times, the starting point for ufology – the study of UFO phenomena – is generally accepted as 24 June 1947.

On this date, Kenneth Arnold (1915-1984)

Left: **Kenneth Arnold points to an illustration showing the general appearance of nine mysterious craft he claimed to have sighted crossing the Cascade Mountains, Washington State, in 1947.** *Bill Rose*

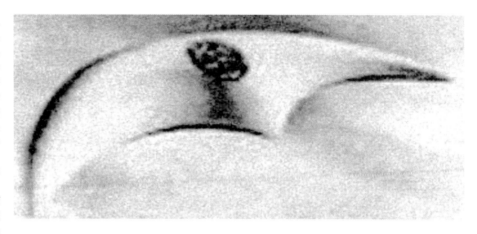

Above: **This illustration shows one of the UFOs allegedly sighted by Kenneth Arnold in 1947. The crescent shape is an artist's interpretation and differs somewhat from Arnold's claim.** Bill Rose

was piloting his light aircraft across an area of the Cascade Mountains in Washington State, searching for the wreckage of a US Marine Corps C-46 transporter that was thought to have crashed in the region. Having decided to abandon the search and head for Yakima, Arnold noticed a formation of nine fast-moving unidentified objects north of Mount Rainer. He made some calculations and determined that their speed was in excess of Mach 2, something far beyond the capability of any existing aircraft.

Initially, Arnold believed he had observed a secret flight of highly classified military planes and he suggested this explanation after landing. But there were no aircraft with this performance and Arnold's report was immediately picked up by the media, who adapted it to imply that the objects might be alien spacecraft.

Newspaper reports also began to spread the term 'flying saucer', which originated with a journalist called Bill Bequette, who mentioned that Arnold had sighted 'nine saucer-like aircraft flying in formation'. Further sightings of unusual aerial objects were reported in the Washington State area, although it has proved impossible to gauge the accuracy or honesty of these accounts. Exactly what Arnold observed from his aircraft remains a complete mystery, but there have been attempts to explain the objects, with the most favoured being a flight of nearby white pelicans, or a train of falling meteor fragments.

Some days later, on 3 July 1947, a New Mexico rancher called William 'Mac' Brazel discovered pieces of unusual debris scattered across one of his fields. He contacted the authorities at nearby Roswell and there were soon reports circulating that wreckage from a crashed flying saucer had been recovered by the military, including several alien bodies. The incident was handled in such an amateurish way by officials that conspiracy theorists now regard this as the biggest cover-up of modern times. However, it is interesting to note that this incident received relatively little attention from serious UFO investigators of that period and several decades passed before the event was elevated into a major conspiracy.

Exactly what took place remains unclear and the Pentagon stuck firmly to its weather balloon story, despite growing public scepticism. Finally, in 1994, Republican Congressman Steven Schiff forced the USAF to re-examine the Roswell Incident. Their new inquiry determined that the balloon account remained the most likely explanation. However, the USAF now suggested that a top-secret 'Project Mogul' balloon had been responsible and they finally acknowledged that

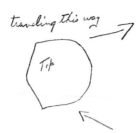

traveling this way

Top

They seemed longer than wide, their
thickness was about 1/20th of their width

Mirror Bright

They did not appear to me to whirl or spin but seemed in fixed position,

traveling as I have made drawing.

a cover-up had taken place.

Project Mogul was a small-scale series of trials to test the feasibility of carrying specialised acoustic equipment to high altitudes where it would listen for evidence of Soviet nuclear explosions. Nevertheless there are many problems with this attempt to explain what happened in July 1947 and few ufologists or conspiracy theorists have accepted it. The true story is uncertain, but it is possible there is some awkward explanation that remains too sensitive to reveal.

Public sightings of unidentified flying objects (UFOs) now gathered considerable momentum and the next major incident took place on 7 January 1948, when a Kentucky Air National Guard Mustang fighter crashed, killing pilot Captain Thomas F. Mantell (1922-1948) after he climbed to investigate a UFO. It is now fairly certain that Mantell had tried to intercept a large, possibly classified, high-altitude balloon and lost consciousness due to oxygen starvation. But the tabloid media capitalised on the incident, suggesting that Mantell's fighter had been shot down with a death ray.

Another event to generate major interest took place during the early evening of 11 May 1950 at a farm near McMinnville, Oregon. Farmer Paul Trent (1917-1998) and his wife Evelyn (1926-1997) claimed to have observed a flying saucer passing over their property and Paul Trent took two pictures of the object with his roll-film camera. By mid-June, the story had reached the national press and the photographs appeared genuine. Nevertheless, recent analysis has cast some doubt on their authenticity.

The next series of important UFO events took place in Washington DC during the second half of July 1952. Numerous unidentified lights were sighted in the night sky above the American capital and radar operators at the airport began tracking unidentified objects that would sometimes accelerate to hypersonic speeds.

Jet fighters were scrambled on several occasions to investigate the supposed intruders, but failed to make contact. What took place during this brief period remains unexplained, but it seems likely that the cause was due to very unusual atmospheric conditions and problems with early radar systems, which were crude by modern standards, easily fooled and often temperamental.

By now, five years of UFO sightings had led to official US interest in the phenomena (although largely public relations motivated) and different opinions had formed to explain the more unusual reports. The general public was leaning in the direction of an alien presence on Earth. The idea of visiting extraterrestrials carried mystery and

excitement, which could easily be exploited by writers and film-makers to enhance the belief.

Whether or not senior Pentagon officials ever seriously considered the possibility that UFOs were extraterrestrial in origin remains unknown, although this does appear to have been an initial concern. But it is now evident from official documents that the US authorities became increasingly worried that flying saucers might be Soviet in origin, perhaps developed from advanced Nazi technology that had fallen into Russian hands at the end of World War 2.

The Cold War was showing no signs of ending and many observers anticipated it would lead to another global conflict, this time involving the use of nuclear weapons. If the Russians were testing long-range supersonic flying discs, perhaps capable of delivering nuclear weapons, it represented a substantial change in the balance of power. Prior to the Kenneth Arnold sighting of 24 June 1947, there had been official interest in unusual lights seen by aircrews during wartime operations (see Chapter 2).

US and British scientists had been given the task of examining these reports and in 1952, some of the same US scientists were recruited by the US Central Intelligence Agency (CIA) to consider the troublesome UFO issue and determine if there was a genuine threat to national security. The committee was headed by Dr Howard P. Robertson (1903-1961), thus becoming the Robertson Panel, who met in secret for the first time on 14 January 1953.

Within months, they concluded that the majority of UFO reports had rational explanations and most were the product of misidentification. But it later emerged that the Robertson Panel recommended that a public relations campaign be initiated to debunk UFOs, and the media used for this purpose. They also suggested that all civilian UFO organisations should be carefully monitored. Apparently, there was concern that the Soviets might use fake UFO reports as a way of masking their activities during an attack on America.

According to UFO investigators, there are still many documents produced by this panel that have never been released and it stands to reason that the CIA recognised the potential of using UFO debunking methods to conceal accidental public observation of highly classified aircraft or unusual operations. By 1954, the Robinson Panel's recommendations had been passed to senior Pentagon officials

A blow-up of the flying disc that was allegedly photographed from the Trent farm at McMinnville, Oregon, on 11 May 1950. NICAP

Flying Saucer Technology

and subsequently a command was issued to all military personnel and commercial pilots that prohibited them from discussing UFO sightings with the media. This directive known as JANAP-146 (Joint Army, Navy and Air Force Publication 146) carried severe penalties for non-compliance.

During the early 1950s, many individuals capitalised on interest in UFOs, with Polish-born George Adamski (1891-1965) taking things to the next level, claiming direct contact with aliens from Venus and other nearby planets. Adamski's blatant fabrications were taken seriously by a surprising number of people and his amateurish photographs of flying saucers proved influential with writers, film-makers and scientists such as Thomas Townsend Brown (1905-1985), who attempted to reproduce the perceived technology.

One writer who took full advantage of interest in UFOs was former Marine Corps officer and pilot Donald Keyhoe (1897-1988). In late 1949, Keyhoe wrote an article called 'The Flying Saucers Are Real' for *True* magazine and this was expanded into a highly successful book the following year. Keyhoe soon established himself as an authority on UFO phenomena and a supporter of the extraterrestrial hypothesis, becoming a co-founder of the National Investigations Committee on Aerial Phenomena (NICAP). Keyhoe was also responsible for inspiring the 1956 Hollywood movie *Earth vs. the Flying Saucers*.

It is now evident that reports of flying saucers were having a definite impact on Western military officials and aircraft designers, who were willing to explore new ideas in an attempt to gain a military advantage and ensure that the Soviets did not steal a massive technological lead.

In Toronto, Canada, the aircraft manufacturer A. V. Roe was developing a top-secret flying saucer fighter. When serious technical problems arose with the project, Pentagon officials stepped in with immediate financial support to keep things running. This would lead to the development of some very advanced proposals that were simply too far ahead of their time to be viable. In addition to this, a number of US defence contractors such as Lockheed were secretly studying high performance long-range supersonic flying discs with a vertical take-off and landing (VTOL) capability.

By the early 1960s, the public's interest in UFOs had declined and designers had largely abandoned plans to build saucer-shaped aircraft, although the US and Russia briefly considered the disc-shaped configuration for spacecraft applications. However, when the Spielberg movie *Close Encounters of the Third Kind* (1977) was released, it generated a significant revival of interest in UFOs. This was eventually followed by the hugely successful SF TV series *The X-Files* (1993-2002) that put ufology firmly back on the map.

Sightings of UFOs around the world rocketed, keeping pace with the show's ratings, and there were growing claims of abduction by small grey aliens. Abductees would be taken onboard spacecraft where bizarre and usually frightening medical experiments were performed on them. Again, these reports began to diminish as ratings for *The X-Files* started to tail off during the final season.

Although this book is not primarily aimed at ufologists or the general UFO community, I have tried to show how flying saucers have been perceived since the Arnold sighting, and the way that some engineers and scientists have been influenced by the world of ufology. I should mention that although a certain amount of fresh information has come to light since my previous publication on this subject, *Flying Saucer Aircraft*, there are still significant gaps in the histories of many military-sponsored projects described in this book. In many instances the (primarily American) documents relating to these projects have simply been destroyed, or lost within a mountain of secret material that nobody can navigate around, or has authorisation to declassify.

Are there ongoing cover-ups? It seems probable and I am fairly certain there are still flying saucer related secrets that remain under wraps. We can only speculate about what these might be. Conceivably, a few circular-shaped prototypes have been built and tested since World War 2. Even if they were total failures, such prototypes remain secret in order to preserve the useful cover that UFO phenomena provide for a range of black projects and operations. On the other hand, it is not my intention to suggest that secretly built circular-winged aircraft can explain most (or for that matter any) of the more reliable, unexplained UFO sightings.

The flying saucer and the pure flying wing or delta remain the most aesthetically pleasing shapes for aircraft, and many of the earliest concepts remain contemporary and futuristic. I suspect that the early designers simply liked the appearance of a circular-winged aircraft and then sought ways to justify it aerodynamically. All early aircraft suffered from a lack of engine power, so maximising lift was essential and it became clear that a circular or oval shape might offer some advantage while retaining a relatively compact shape.

Below: **A photomontage based on many claimed UFO sightings above Washington DC during July 1952. This highly publicised and very intensive period of UFO reports from around the capital would become known as 'The Washington Flap'.** Bill Rose

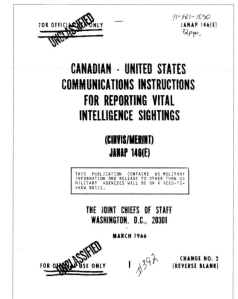

Left: **This mysterious object was allegedly photographed above Rouen, France, in 1957 by a French air force pilot. It is surprisingly similar to the McMinnville saucer, but the lack of technical detail regarding the sighting strongly hints at this being a fake, possibly based directly on the McMinnville image.** *Bill Rose*

Below left:**This drawing shows the general appearance of an Adamski flying saucer from the early 1950s that eventually influenced science fiction film-makers and aircraft designers.** *Bill Rose*

Below: **This Pentagon directive known as JANAP-146(E) was one of a series of documents prohibiting unauthorised disclosure of UFO sightings.** US DoD

When moving from a broad ring to an enclosed circular wing, designers recognised the scope for maximising the use of internal space, and by the middle of the last century some engineers could see the potential for a fully enclosed propulsion system that was specifically designed for a circular-shaped aircraft. If this system could be used to provide VTOL operations, the aircraft was a potential winner.

At higher speeds, the large surface area of a flying disc can create drag issues, but it was evident that a disc-shaped aircraft with suitable propulsion might be capable of high supersonic cruise at substantial altitudes. Designers also recognised that the traditional flying saucer shape would make a very good low visibility 'stealth' aircraft, leading to claims that the US has built a small number of highly classified disc-shaped spyplanes. These would be totally deniable because of their UFO appearance and casual observers could easily be discredited.

Nevertheless, there are recognisable problems with the early disc-shaped proposals. Visibility would be a problem for any centrally positioned pilot, engine failure would probably result in the aircraft becoming unflyable, and flight control could prove difficult. These factors and many other technical considerations go some way to explaining why there are no flying discs in our skies at the present time.

In this book, I have briefly outlined modern interest in disc-shaped aircraft, the connection to unknown phenomena and science fiction. However, the circular-winged flying machine has its roots much further back in time, even before the first manned flights took place. In the following chapters, I will be looking at the overall history of circular-shaped, (mostly) manned aircraft, from early concepts to lighter-than-air craft.

It is not my purpose to ridicule ufology, which remains a fascinating subject. I have tried hard to deal with this topic in a commonsense, objective manner and avoid any of the wild speculation found on the internet about Man-Made-UFOs (MMUFO). Only a small number of the aerial vehicles featured in this book can be described as true flying saucers and many fit the description 'discplane' somewhat better.

I have reviewed many craft from VTOL spade and heel shaped fighters to high-altitude balloons associated with UFOs. All of the projects described in this book are linked in various ways to the general theme of man-made flying saucers. Hopefully, I have shed some light on a small number of unusual aircraft that have frequently been dismissed as non-existent fantasy concepts.

Bill Rose
Norfolk, England, July 2011

Acknowledgements

Many people have assisted with the production of this book; I would especially like to thank Chris Gibson, Joel Carpenter, Robert Bradley, David Myrha, Martin Müller, Alexi Malinovsky, Tony Frost, Andrew Gibbs, John Gibbs and Jack M. Jones.

Glossary

AAM: Air-to-Air Missile

AFB: Air Force Base

Aspect Ratio: The aspect of wingspan to mean chord, usually calculated by dividing the square of the span by the wing area

BAC: British Aircraft Corporation

Black Projects: Highly classified, secretly-funded programmes usually involving the development of new military systems. These can remain hidden for years, perhaps decades, sometimes progressing no further than studies

BLC: Boundary Layer Control, a method of drag reduction

BTZ: Bureau Technique Zborowski. An aircraft design organisation, operating in post-war France and headed by Dr Helmut Zborowski

CAA: Civil Aeronautics Agency (US); Civil Aviation Authority (UK)

CDRB: Canadian Defence Research Board

CIA: Central Intelligence Agency

Coanda Effect: The tendency of a jet stream to flow along a solid surface, which may curve away from the jet axis

DARPA: US Defense Advanced Research Projects Agency

DFS: Deutsche Forschungsanstalt fur Segelflug – German Research Institute for Soaring Flight

DOD: Department of Defense

Ducted Fan Design: Enclosed rotor propulsion system

FAA: US Federal Aviation Authority
GETO: Ground Effect Take-off. A rolling take-off, initially using ground effect lift

L/D: Lift/drag ratio. A measurement of aerodynamic efficiency, with a maximum of about 45:1 for sailplanes

Luftwaffe: German Air Force

Mach Number: Ratio of an air vehicle's true speed to the speed of sound in air at the altitude the vehicle is flying. A vehicle travelling at a Mach number in excess of 1 is considered to be supersonic. At velocities above Mach 5, speed becomes hypersonic

MoD: UK Ministry of Defence, 1964-present day

MIT: Massachusetts Institute of Technology

Mt: Megaton

NACA: National Advisory Committee on Aeronautics – the forerunner of NASA

NASA: National Aeronautics and Space Administration, created on 1 October 1958 by the Eisenhower Administration

NAVAIR: US Naval Air Systems Command

NGTE: UK National Gas Turbine Establishment

OKB: Opytnoe Konstruktorskoe Byuro – Soviet Experimental Design Bureau

Payload: Normally cargo or equipment, but can refer to military ordnance

RAE: Royal Aircraft Establishment

RAND Corporation: Originally Project RAND (Research ANd Development) was established in 1946 as a branch of Douglas Aircraft Company in California. In 1948, RAND separated from Douglas, becoming a non-profit-making think-tank

Rawin: Derived from 'Radar wind'. A radar reflective target used with balloons to assist tracking

RCAF: Royal Canadian Air Force

RFGT: Radial Flow Gas Turbine. An unusual high performance pancake-shaped engine, working edge-on during level flight

RLM: Reichsluftfahrtministerium (World War 2 German Air Ministry)

RRG: Rhön-Rossitten Gesellschaft – early German gliding society. In 1933, RRG became the DFS

Stealth: Low-observable technologies

STOL: Short Take-Off and Landing

T/W: Thrust to weight ratio

Tons/tonnes: Throughout this book the value for 'ton' equates to an American short ton, which equals 2,000lb. The metric tonne is 1000kg, and the conversion factor is 0.9072

UAV: Unmanned Aerial Vehicle

USAAF: United States Army Air Force

USAF: United States Air Force

Volksjäger: German wartime People's Jet Fighter

VTO: Vertical Take-Off

VTOL: Vertical Take-Off and Landing

WSMR: White Sands Missile Range

Chapter One

Early Designs

The earliest recorded use of a circular spinning disc was at athletic events, and the practice of throwing a bronze discus became an established part of the Greek Olympic Games by 708BC. Jugglers also used the gyroscopic stability of spinning plates for entertainment purposes, although the origins are unclear and equally old.

In the late 19th century, the discus evolved into a cheap toy called the Frisbie when American college students began using pie plates for unapproved games. Pies were supplied to most New England colleges by The Frisbie Baking Company (1871-1958) of Bridgeport and the word Frisbie was stamped into all their metal plates.

In 1948, the Frisbie became a legitimate toy when Walter Frederick Morrison and Warren Franscioni devised a more aerodynamic plastic version of the spinning plate, which capitalised on growing interest in UFOs. After the rights to the Frisbie had been sold to a larger concern, the name was changed to Frisbee, which became a trademark currently owned by the toy manufacturer Mattel Inc. The US Navy also studied the properties of the Frisbee during the late 1960s, when they attempted to devise an effective flare dispenser.

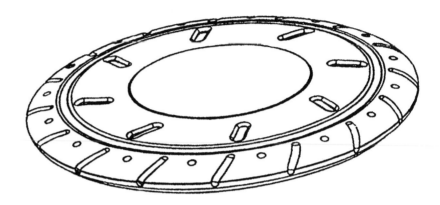

SWEDENBORG

Aside from stabilised spinning plates, the idea of a circular-shaped flying apparatus dates back quite a long way. In the early 18th century, a Swedish theologian, philosopher and scientist called Emanuel Swedenborg (1688-1772) drew up plans for a manned flying device with a circular or elliptical wing. Swedenborg finalised his study in September 1714 and he intended to utilise flapping wings for propulsion and directional control. This one-man contrivance would be built almost entirely from wood and covered with sailcloth.

The design was completely unworkable, but the flying machine was featured in

Sweden's first scientific journal, *Daedalus Hyperboreus*, during 1716. The circular flying device would have a diameter of 25ft (7.62m), and an oval alternative would measure 32ft by 24ft (9.75m by 7.3m). Swedenborg intended to start with a scale-

Above: **This is the current design patent drawing for the world's most successful flying saucer, the Mattel flying disc toy.** *US Patent Office*

Below left: **Emanuel Swedenborg, the 18th century Swedish scientist, philosopher and mystic, who designed an oval-shaped flying machine in 1714.** *Bill Rose*

Below: **Original drawing of Swedenborg's flying apparatus, as shown in his notes of 1714.** *Bill Rose*

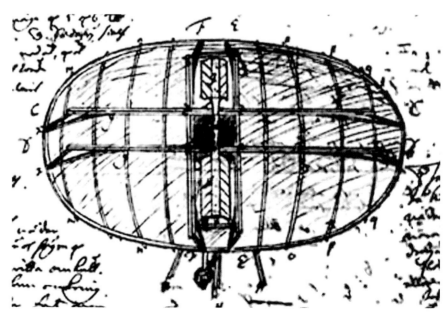

sized model, but there were few supporters of his project and no financial backing for an idea considered to be bordering on lunacy.

Swedenborg's flying apparatus was soon forgotten, but the idea of manned flight was already established thanks to Fr Francesco Lana-Terzi (1631-1687). He was a Jesuit priest and a professor of mathematics at Brescia, Italy, who suggested a way to fly in 1670.

Unfortunately, his ideas for using metallic vacuum balloons were unworkable and more than a century passed before the Montgolfier Brothers accidentally discovered the principle of a hot air balloon. Subsequently, the first manned flight was undertaken on 21 November 1783 in Paris. By now, the British had discovered how to produce hydrogen and this led to gas-filled balloons, which closely trailed the Montgolfier Brothers' work.

PÉNAUD'S AIRCRAFT

Many decades later, the brilliant French inventor Alphonse Pénaud (1850-1880) built a small hand-launched model aircraft with an 18in (457mm) wingspan and an overall length of 20in (508mm). It was powered by a rear-mounted 8in (200mm) propeller that was driven by a rubber band, and Pénaud undertook a demonstration at the Tuileries Gardens in Paris on 18 August 1871. This amazingly advanced model covered a distance of 131ft (40m) in eleven seconds and led Pénaud to design and build several rubber band-powered helicopters using contra-rotating propellers.

Below:**Alphonse Pénaud, the aeronautical designer who was responsible for the first demonstration of powered flight and the creator of advanced designs many decades ahead of their time.** *Bill Rose*

Pénaud now started work on the design of a full-sized manned flying machine, producing an elliptically shaped seaplane powered by two propellers connected to a steam engine. Having secured a design award from the French Academy of Sciences in 1875, Pénaud then teamed up with an engineer called Paul Gauchot. This partnership led to a very advanced design for an aircraft, based on Pénaud's earlier work, which was patented in France during 1876, receiving the (now defunct) reference 111574.

This elliptically shaped aircraft would be propelled by two forward-positioned four-bladed metal propellers with variable pitch. The unresolved problem with this design was obtaining a suitable powerplant to drive the propellers. However, Pénaud was hopeful that a lightweight steam engine would be developed with sufficient power to allow a level flight speed of approximately 60mph (96kph).

The Pénaud-Gauchot aircraft would carry a crew of two. It was a masterclass in design innovation, utilising a retractable four-wheel undercarriage, a glass-covered cockpit with

flight instruments and a control stick in the cockpit linked to a rudder and ailerons. There were flaps and the wing was to be covered in varnished silk. Although upper and lower stays were planned, it was hoped that these could be dispensed with when subsequent models were built. Proposed dimensions and weights remain unclear.

In 1880, Pénaud attempted to gain the support of airship pioneer Henri Giffard. Unfortunately he dismissed Pénaud's ideas as unworkable and, as a direct consequence, Pénaud, who appears to have been bipolar, committed suicide. Whether or not the Pénaud-Gauchot aircraft would have flown with a suitable engine remains unknown, but this design was decades ahead of its time and proved very influential.

Inventors produced numerous concepts for circular and elliptical flying machines during the last part of the 19th century that included platforms suspended beneath balloons and gliders. The best-known glider pioneer during this time was the German inventor Otto Lilienthal (1848-1896). He

The extraordinarily advanced Pénaud-Gauchot steam-powered aircraft design of 1876. *Bill Rose*

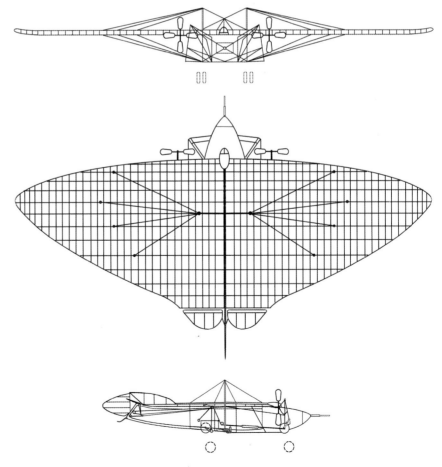

Left: **A replica of the Ezekiel flying machine housed at the Northeast Texas Rural Heritage Museum in Pittsburg, Texas.** *Northeast Texas Rural Heritage Museum*

clearly demonstrated that heavier-than-air machines could be flown without the need for flapping wings and established the viability of aircraft with unconventional shapes. Like many of his colleagues, Lilienthal pushed the boundaries of his research too far and died on 10 August 1896, following a glider crash.

EZEKIEL

Possibly the most interesting (and little known) pioneer of this period was the American Baptist Minister, Rev Burrell Cannon (1848-1922), who may have been responsible for the world's first powered (although unacknowledged) flight. As a skilled engineer, Cannon began to design a powered flying machine in 1898 and once he secured the financial support of several friends, construction began in P. W. Thorsell's Foundry at Pittsburg, Texas. Many parts were custom built and most of the assembly was carried out by an engineer called Gus Stamps, who completed the job in August 1901.

On completion, Cannon named the semi-elliptical aircraft 'Ezekiel'. It had a span of 26ft (7.92m) and was built mainly from wood, covered by canvas. For propulsion, Ezekiel was powered by a small 4-cylinder internal combustion engine that appears to have been rated at 80hp (59.6kW). It was hand built in the foundry and was used to drive four separate wheels with paddles that behaved rather like propellers. Cannon patented his propeller wheels, considered the most important feature of the project, in 1901.

Once the aircraft was completed, it was moved outside the factory, probably for engine testing. Ezekiel remained there until some

unrecorded date in late 1902, when Stamps undertook an unauthorised test flight without Cannon present. Exactly why this happened remains unknown; it may have started out as a taxiing trial. However, it appears that the aircraft left the ground, attaining a height of approximately 12ft (3.65m) and travelled for a distance of about 160ft (48m).

Apparently, Stamps finally flew across a fence into an adjoining field before making a hard landing, which damaged the undercarriage. Assuming this is true, it rewrites aviation history, becoming the world's first powered flight. But, there was no proper documentation of the event, the aircraft was damaged and a lack of funds held up repairs and prevented any further attempts.

When the damage to Ezekiel was finally fixed in early 1903, Cannon decided to ship the aircraft by rail to St Louis for the World Fair.

(This event was scheduled for late 1903, but was postponed until 30 April 1904.) While en route to St Louis, a fierce storm blew the aircraft from the freight platform near Texarkana, Texas, and it was completely destroyed. Apparently, a replica was eventually built in Chicago, Illinois, and flown in 1913, only to be badly damaged on its first flight attempt. At this point, Cannon abandoned any further attempts to produce a flying machine. Although the details of Cannon's project were largely forgotten, Bob Loughery, a Texan engineer, built a replica of the aircraft in 1986, which is currently on display at the Northeast Texas Rural Heritage Museum in Pittsburg, Texas.

Although Cannon's unusual aircraft had failed to become the world's first powered and manned aircraft to fly, the competition to achieve this honour was fierce. Another American inventor, Lyman Wiswell Gilmore Jr, claimed to have built and flown a steam-powered monoplane in California during May 1902, but again, there is no hard evidence to support this.

On 17 December 1903, Orville Wright (1871-1948) made a 12-second flight at Kitty Hawk, North Carolina. He flew for a distance of 120ft (36.5m), securing a place in history for him and his brother Wilbur (1867-1912). Interestingly, both the Wright Brothers said that a model aircraft designed by Alphonse Pénaud had inspired them.

Below: **John Kitchen's annular-winged biplane at Middleton Sands, circa 1911.** *Bill Rose*

Flying Saucer Technology

BRITISH DESIGNS

The early years of manned flight saw rapid technical advances, but the need for structural strength and maximum lift led to the widespread adoption of one or more wings stacked above another. While Louis Blériot (1872-1936) managed to build the world's first successful powered monoplane in 1907, the biplane remained dominant for many years. However, some designers wondered if circular-shaped wings could be used to improve biplane performance by providing more lift. It seemed possible that as engine performance improved, this design would allow the upper wing to be dispensed with entirely, thus reducing drag and weight.

Less than six years after the first manned flight, the pace of development was beginning to gather momentum. Louis Blériot crossed the English Channel on 25 July 1909, the Wright Brothers toured Europe and the next major advance was expected to be a commercial aircraft. One constructor hoping to cash in on this development was Captain Joseph Donovan, who established the Donovan Aeroplane Company at 11 Church Street, West Hartlepool, England, in May 1909. With two associates and a working capital of £2,000, Donovan had already started to construct an aircraft that used circular wings. The assembly was completed by October 1909 and Donovan's flying machine was ready for testing.

The unnamed aircraft had a span of 28ft (8.5m), although the length is unknown. A pair of two-blade pusher propellers provided propulsion, supplemented by two horizontal propellers intended to generate extra vertical lift. A 30hp (22kW) 6-cylinder Fothergill engine generated power for this complicated system, but with the aircraft's weight estimated at 1,000lb (454kg), it was hopelessly inadequate. Once taxiing trials commenced, the lack of power became all too apparent. The aircraft was barely able to move under its own weight and consequently the project was abandoned.

Another aviation pioneer was the Lancashire-based inventor and engineer John George Aulsebrook Kitchen (1869-1940). He filed numerous patents for innovative improvements to tyres, carburettors and new detergent mixtures, finally turning his attention to manned flight in 1910. This soon led to the construction of a biplane with circular wings.

His flying machine was largely completed by the following year and built almost entirely from wood. The fuselage was an open framework and the wings were covered with fabric. Propulsion was provided by a two-

blade wooden propeller mounted at the front of the aircraft, which was driven by a 7-cylinder 50hp (37.28kW) Gnome Omega rotary internal combustion engine. The pilot sat directly behind, which was far from pleasant due to noise and the flow of exhaust gas from the engine, which contained unburned castor oil (usually Castrol R). However, I should add that this problem was far from unusual in the early days of powered flight.

The fixed undercarriage consisted of twin skids with two pairs of spoked wheels at the front and small runners at the rear of the aircraft. Dimensions, weight and estimated performance are unknown, although it is possible to make rough estimates. The aircraft was largely completed by early 1911, but Kitchen lost interest in the project and the aircraft was stored away in a large shed at Middleton Sands, near Heysham, Lancashire. Nevertheless, Kitchen continued to design aircraft for the next year or so, producing plans for more advanced monoplane designs with semicircular and parabola-shaped wings.

Kitchen's biplane was then sold to Cedric Lee (1882-1916), who was the wealthy owner of a weaving business. Lee now sought the assistance of his friend George Tilghman Richards (1885-1960), who was also a designer and engineer. The two men completed work on the aircraft around June 1911 and the biplane was initially tested as a glider. Cedric Lee piloted the aircraft, which he reported to be very stable in flight, while Richards made, or organised, a series of ongoing modifications to leading edges and various other parts of the aircraft.

Although these trials went well, the aircraft was damaged in autumn 1911, after running into a small gully while landing and it was then kept in the shed at Middleton Sands while awaiting repairs. Unfortunately, the shed collapsed during a severe storm in November 1911 and the aircraft was destroyed. It is not clear if any components were salvaged from the wreckage, but Lee and Richards considered this to be a setback rather than a complete disaster, having already decided to build two model gliders that would hopefully allow them to develop a perfect leading edge camber.

The two men devoted plenty of time to this project and must have had assistance, because the models were completed by the beginning of 1912. Each had an overall span of 4ft 8in (1.42m) but it is unclear if there was any significant difference between them.

Testing took place within a large empty hangar and progressed well, with it becoming apparent that extra camber produced a flatter glide angle. As a direct result of these trials,

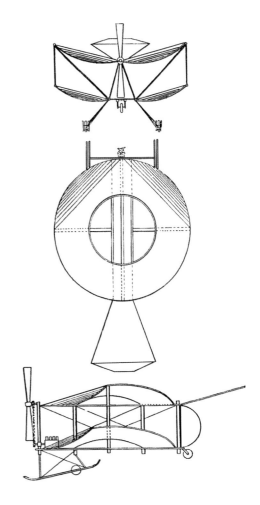

Above: **One of several pre-World War 1 circular-winged aircraft designed by John Kitchen.** *Bill Rose*

Below: **Portrait of wealthy businessman Cedric Lee.** *Bill Rose*

Lee decided to fund the construction of a second manned aircraft. It would be an unpowered glider, with an overall span of 22ft (6.7m) and an upper wing area of 400sq ft (37.17m2). The empty weight of this aircraft is recorded at 215lb (97.5kg), rising to 710lb (322kg) with the pilot and some ballast. They must have been working on this glider alongside the models, as it was ready for testing in February 1912.

Trials were undertaken at Sellet Banks, alongside the River Lune between Whittington and Kirkby Lonsdale. Lee made most of the flights, with the glider being launched by means of a catapult and track system, involving the use of a large weight inside a tripod frame. He found the aircraft very stable in flight and easy to control, although it was decided to add extra control surfaces as the project progressed.

Some film was shot of the trials and a brief clip of Lee making a test flight was shown at the start of Ken Annakin's 1965 movie *Those Magnificent Men in their Flying Machines*. In addition, a fairly accurate non-flying replica of the original powered aircraft was built by Denton Partners at Woodley for use in this production. It was seen to be flown in the film by a character called Harry Popperwell, who was played by the British comedian Tony Hancock (1924-1968). This recently renovated aircraft is now on display at the Newark Air Museum, Nottinghamshire.

In December 1912, the test flights at Sellet Banks were concluded. Richards was now making regular trips down to the East London College, where he set up a small 2ft (0.6m) wind tunnel to test models with various flat annular wing profiles. By January 1913, Lee and Richards had fully relocated to London where a business called Cedric Lee Co was established.

They had been drawing up plans for a more advanced aircraft that may have been inspired by Kitchen's later proposals. Once the blueprints were completed, Lee hired an engineer by the name of James Radley and his assistant Eric England to construct the new aircraft at Shoreham Aerodrome in Sussex.

Top: **A replica of the original Lee-Richards annular-winged biplane built by Denton Partners at Woodley in 1964 and used in the filming of *Those Magnificent Men in their Flying Machines*. Restored at Shoreham at the turn of the century, this replica is now on display at Newark Air Museum.** *Chris Gibson*

Middle: **The Lee-Richards No 1 monoplane at an advanced stage of construction.** *Bill Rose*

Bottom: **Prior to completion, a front view of the No 1 Lee-Richards monoplane.** *Bill Rose*

Although this was not their first aircraft, it was decided to call it the Lee-Richards No 1 monoplane, but this unusual design had soon become known as the Circleplane. Accommodation was provided for the pilot and an observer, and the wingspan is understood to have been 20ft (6m), with a nominally greater length. The exact height is uncertain. Steel cables were attached to an upright forward support and used to brace the wing. The mainly wooden fuselage and wing area was covered in fabric. Propulsion took the form of a forward-positioned two-blade wooden propeller, driven by a 7-cylinder 80hp (59.6kW) Gnome Lambda rotary engine. The fixed undercarriage comprised two main wheels with two smaller forward in what was almost a tricycle arrangement. The exact weight of the aircraft remains unclear at present.

No 1 was completed in late 1913 at Shoreham and flown before the end of the year by test pilot (Eric Cecil) Gordon England (1892-1976), who took off without undertaking any taxiing. The aircraft performed well, showing excellent lift and good handling. Gordon England flew it around a 5-mile (8km) circuit, reaching a speed in excess of 70mph (112kph). On his landing approach, it appears that England used his blip switch to interrupt the ignition and reduce speed, but may have been unable to restart the engine. There was also a weight distribution problem making the aircraft tail-heavy and this combination resulted in a very hard landing.

No 1 was badly damaged, although England escaped with little more than a few cuts and bruises. But it was possible to salvage some parts of the aircraft and the assembly of a replacement monoplane (No 2) was under way in early 1914.

Richards also took the opportunity to make a number of small improvements to the design, the most noticeable being a change to the control surfaces. Lee-Richards No 2 was completed in March 1914 and England took the aircraft on its first test flight, reporting that it suffered rather badly from yaw problems.

Charles Gordon Bell (1889-1918) replaced England as the test pilot and as trials continued, further modifications were made to the control surfaces. Flights continued until April 1914, when Bell lost control of the aircraft after an eyebolt came away from one of the elevators at about 800ft (244m) altitude, jamming it in a 'hard down' position. The elevator then fully detached while Bell was attempting to land and the aircraft pancaked.

This resulted in No 2 being damaged beyond repair, although Bell escaped with minor injuries. Nevertheless, some useful components were recovered from the wreck, including the valuable Gnome engine and the propeller extension shaft. It was reported that two further aircraft were under construction at this point, with plans to enter them for the Gordon-Bennett Aviation Race, which was due to be held in France during September or October 1914. So priority was given to completing one of these aircraft, which received the designation Lee-Richards No 3.

Visually similar to the earlier designs, this aircraft incorporated various further improvements. Bell undertook many test flights in No 3 and demonstrated it to Winston Churchill and Field Marshal French on one occasion. Most of the bugs had now been eliminated and Bell continued to fly No 3 until the start of World War 1. Bell was then called up to serve in the Royal Flying Corps so Lee took over as the test pilot. Unfortunately, he managed to crash this aircraft into the River Adur near Shoreham Aerodrome. The Gordon-Bennett Aviation Race had been scrapped and this accident spelt the end of Lee's attempts to create a new type of aircraft with commercial potential.

Soon after the outbreak of World War 1, Cedric Lee joined the Royal Naval Volunteer Reserve (RNVR) as a junior officer and was killed in action during November 1916 at Beaumont Hamel in France. In 1918, Bell was killed while test flying the prototype Vickers F.B.16E at Villacoublay in France. George Tilghman Richards MBE remained in aircraft engineering, becoming a senior member of the Royal Aeronautical Society (RAeS). During a RAeS meeting in August 1948, Richards gave a presentation on his past work which was accompanied by some film of the Lee-Richards aircraft. Apparently, he could

Lee-Richards No 3 circular-winged monoplane in flight. *Bill Rose*

not resist making the remark that their machine was "the genuine and original flying saucer!" He died in 1960.

In 1997, an Australian aviation enthusiast called Ron J. Feast began work on what might be described as an updated version of the Lee-Richards Circleplane. Mainly built from wood, his aircraft had a wingspan of 23ft (7m) and was powered by a forward-mounted Rotax 912 UL motor providing 80hp (59.6kW) and driving a two-blade wooden propeller. The aircraft was test-flown by Graham White at Camden Airport in August 2000. Trials continued for a year and a number of changes were made to the aerodynamics including new wing sections, which decreased the span to 20ft (6m). The undercarriage was also converted from two forward main wheels and a tail wheel to a new tricycle layout. In 2003, the Mk 2 Feast aircraft was completed, with tests beginning in 2005.

THE UMBRELLAPLANE

In 1910, the New York-based inventor William Swart Romme built an unusual model aircraft that won him first prize in a design contest. The circular-shaped model was apparently very stable in flight and considered to be an original concept. Romme hoped to find backing for a powered, full-sized, man-carrying version and applied for a patent on 14 March 1910. Romme's design soon gained the attention of wealthy Chicago industrialist Harold Fowler McCormick (1872-1941), who agreed to fund development of the flying machine.

It was McCormick's intention to sidestep the patents already established by the Wright Brothers with this unusual aircraft and develop the design for commercial purposes.

Work on the first aircraft began at Mineola, Long Island, New York, in summer 1910. It is unclear if the plane was being kept outdoors at this time, but it was badly damaged during a storm in November 1910. The operation was then moved to a new workshop at Belmont, New York. Some parts were reused and the one-man, circular-shaped aircraft had been completed by February 1911.

A 7-cylinder 50hp (37.28kW) Gnome Omega rotary internal combustion engine was chosen for propulsion, positioned towards the centre of the aircraft to maintain the centre of gravity and connected to a wooden two-blade pusher propeller at the rear by a long drive shaft. The overall span of the aircraft was about 35ft (10.6m) and the fabric-covered circular-shaped wing was supported by steel cables attached to a central upright post fixed to the fuselage section. A number of different fixed position undercarriages were tested on this aircraft, with the initial design consisting of two widely spaced forward spoked wheels and a tail wheel. This arrangement was reversed and replaced with several combinations of skids and wheels more closely positioned to the fuselage section. There were also a variety of different aileron and rudder designs tested. When Romme originally conceived the aircraft, he considered the possibility of attaching gas bags as a weight reduction measure, but this idea appears to have been abandoned. The gross weight of the aircraft is uncertain, although it is thought to have been relatively low. There was no cockpit as such,

just a seat directly ahead of the engine and a steering wheel arrangement, connected to the control surfaces.

There is some conflicting information about the development history of this machine, but it seems that some engine testing was undertaken at Belmont and then a decision was made to disassemble the aircraft and ship it to San Antonio, Texas. This was considered to be a good place to conduct flight trials, as the weather was fairly consistent.

Romme undertook taxiing trials but the aircraft ran over some rough ground and flipped over. He was unhurt but the plane was quite badly damaged. The team then relocated to The Aero Club of Illinois' new airfield at Cicero, Chicago. Better engineering support was available and McCormick was one of the Aero Club's vice presidents, owning the land they used. The aircraft had been called the Cycloplane, although the name Umbrellaplane came into widespread

Right: **A portrait of 'Chance' Vought who assisted McCormick and Romme with the development of the Umbrellaplane.** *Vought Aviation*

Below: **The very distinctive Umbrellaplane.** *Bill Rose*

Above: **Harold F. McCormick and William S. Romme's Umbrellaplane undergoing preparation for flight. The exact date of this photograph is unknown.**
Bill Rose

use after Aero magazine used it to describe the project.

McCormick was now funding development of the Umbrellaplane and another experimental aircraft, which placed considerable demands on the team. As a consequence, McCormick hired Sidney V. James who was a college graduate and Chauncey 'Chance' Milton Vought (1890-1930), who was a talented engineer originating from Long Island, New York. Vought would eventually become a major figure in the US aviation industry.

Yet another part-time member of McCormick's expanding group was the well-known pilot Andrew Drew (1885-1913). He was already employed as the General Manager of Cicero Flying Field, receiving a basic salary of $40 per week, which was good money for that time. The Cicero Flying Field

was formally opened on 4 July 1911 and work continued to prepare the Umbrellaplane for its maiden flight. Many further modifications had been made to the ailerons, controls and undercarriage. Finally, on 23 August 1911, Romme made a very brief test flight, attaining a height of 15ft (4.57m).

However, it was not possible to improve on this performance and it was clear that the aircraft lacked sufficient engine power. Vought suggested a number of further modifications that included moving the pilot and engine slightly rearward. But the problems persisted and it was decided to move the propeller to a forward position. Vought suggested further changes to the wing profile and during April 1912, André Ruel was able to gain slightly more altitude during the next test flight. Throughout the remainder of the year, further modifications were undertaken and it is believed that at least seven major design revisions/rebuilds were undertaken during this project. Brief lift-offs were achieved before the aircraft was put into storage during the winter.

Trials resumed in March 1913 and a number

of short flights were made. But the design was flawed from the outset and McCormick lost interest in the project, deciding to call a halt to any further development. Apparently, he had spent at least $50,000 on the Umbrellaplane, which was a colossal amount of money at that time. The fate of the aircraft is unknown, but it is said that that the structure was broken up and dumped in Lake Michigan.

WEBB'S FLYING MACHINE

During the final year of World War 1, Chicago-based inventor John W. Webb turned his attention to devising a new type of aircraft capable of VTOL and hover at altitude. His aircraft was primarily intended for military use as a light bomber, surveillance or airborne radio platform, with the possibility of producing a naval version configured to attack enemy submarines. The design was completed by October 1918 and it had an almost circular shape with spars radiating from a central post. The aircraft would primarily be built from wood with some metal panels, with braces and cables visible around the periphery. A fully enclosed forward-positioned cockpit seems large enough to accommodate two to four crew members.

The undercarriage was quite unusual, comprising four flexible rubber cylinders inflated with air that would allow soft

Below: **General internal layout of the Webb flying machine, showing the four engines, blowers, cockpit and control system.** *US Patent Office*

Below left: **This extraordinary circular flying machine was designed by John W. Webb in 1918. It was intended for military use, and in addition to possessing a full VTOL capability this aircraft would be capable of maintaining a fixed position at altitude during surveillance and radio relay operations. Whether it could have been made to actually fly is another matter.** *US Patent Office*

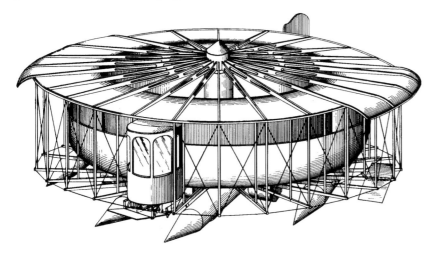

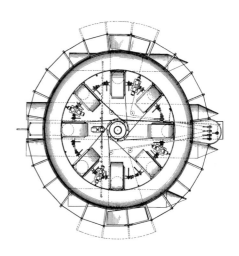

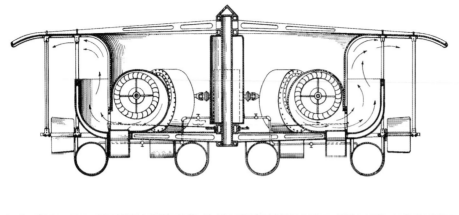

touchdowns and the possibility of landing on calm water. For propulsion, the Webb aircraft used eight centrifugal fans drawing air from above. This would be directed downwards for VTOL and hover, or horizontally for level flight. The fans would be driven by four rotary engines with an anticipated total output of 400hp (298kW). To control the aircraft in flight, it would be possible to regulate the output of the fans using a valve and deflector box system, and there would be ailerons and a rudder.

No figures were ever quoted for anticipated weights or performance, although it seems very unlikely that this design could have been made to work. Nevertheless, it contained a number of interesting features that would eventually be utilised in VTOL aircraft, and Webb's Flying Machine can be seen as something of an early ancestor to concepts produced by designers such as Weygers and Wibault.

THE ROUNDWING

Steven Paul Nemeth worked as a flying instructor at McCook Field, Ohio, during the 1920s. While there, he began to develop ideas for a new type of aircraft that would be easy to fly, comparatively safe, and cheap to build. He started drawing up plans in 1926 and then built a scale model of his design, which was tested in the University of Michigan's wind tunnel.

The results were encouraging and after making further changes to the design, Nemeth funded the construction of a full-size aircraft, which was undertaken by students at Miami University, Ohio. The aircraft was completed in 1934 and Nemeth called it the Roundwing, but rather confusingly it was also called the Umbrellaplane, despite having nothing in common with the earlier McCormick-Romme flying machine.

The Roundwing's fuselage was built from a slightly modified two-seat Alliance A-1 Argo. The original tailplane assembly was retained, along with the fixed undercarriage that consisted of two forward wheels on struts and a tail skid. It is likely that Nemeth obtained this fuselage at a bargain price after Alliance Aircraft of Ohio ran into serious financial trouble following the 1929 Wall Street Crash and were forced out of business, having completed only twenty planes.

The overall length of the Nemeth aircraft was 20ft (6m) and it carried a single overhead circular wing with a span of 16ft (4.8m). Large control surfaces formed the trailing edge, and these were revised several times during the aircraft's life. A forward-mounted 90hp (67kW) Lambert engine drove a two-bladed wooden propeller, although the original A-1 Argo had been designed to take a more powerful radial engine.

The Roundwing was assigned the registration X13651 and was flown on many occasions from Curtiss Airport during the next two years. Various minor modifications were made to the aircraft and it featured in several magazine articles, being variously described as the 'Saucer plane', 'Parasol' and 'Parachute Plane'.

Nemeth did his best to sell the concept on the grounds of safety and exceptional Short Take Off & Landing (STOL) performance, demonstrating how the aircraft could lift off in 63ft (19m), climb at an angle of 45° and land within 25ft (7.6m) at only 30mph (48kph). He also suggested on at least one occasion that the aircraft could be flown by a complete novice with just 30 minutes of instruction! This appears to have been unconsidered sales talk as opposed to common sense. Flight-testing continued into 1936, with a more powerful air-cooled 7-cylinder 125hp (93.2kW) Warner Scarab 50 engine being installed. This provided a maximum speed of about 140mph (225kph) in level flight.

Having proven that a completely circular wing could provide excellent stability, resistance to stalling and the ability to be stored in a garage-sized hangar, Nemeth hoped to start manufacturing the Roundwing for $1,400 per copy. But money remained tight, there was probably some resistance to the unorthodox shape, and his plans were finally abandoned. The fate of the Roundwing is unknown, but it is thought to have been scrapped.

JOHNSON'S UNIPLANE

From an early age, Richard Burton Johnson (1907-2006) was interested in aviation and he finally decided to take an evening course in aeronautics at the Chicago Aeronautical University during 1931. Johnson was fascinated by the idea of developing an aircraft with STOL or ideally full VTOL performance and was encouraged to create just such a design for part of his coursework.

Johnson's first idea was a circular-winged aircraft, which used a central rotor to generate lift, and he called this the Centricopter. He is believed to have built a small model of this design, but soon found the idea was unworkable. So he reconsidered the concept, deciding to concentrate on STOL and the use of a circular wing to provide adequate lift. His design continued to evolve and on 12 August 1931, Johnson was confident enough to file a US patent, which was published the following year (1887411).

He remained hopeful that a manufacturer would be found to build his aircraft, but these were difficult times and nobody was willing to take such a risk. The only alternative was to do the work himself and over a period of several months Johnson managed to raise about $1,000 for the project. His next step was to secure a complete fuselage section and engine from the locally based James Church Airplane Co, which produced very elegant sports monoplanes as

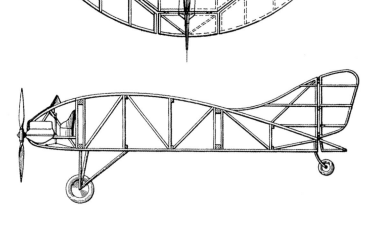

Left: **This drawing is based on a US patent filed in August 1931 for a circular-winged monoplane by Richard Johnson. The concept was flawed from the outset, but Johnson built and tested the aircraft.** *Bill Rose*

'home build' kits. These components were then delivered to his parents' Chicago home at nearby Wabash Avenue, where he began assembly work in the garage.

His aircraft was now called the Uniplane and the most difficult part of the project was building the new circular wing. Various modifications were made to the original design, the most significant being the use of two tail fins, fitted with an elevator between them. The original Church undercarriage was retained, consisting of two forward wheels and a tail skid. When it was fully assembled, the Uniplane measured 16ft 6in (5m) in length and the new wing had an overall span of 14ft (4.26m) and a central thickness of 3ft (914mm). The Uniplane appears to have been a single-seater, but it may have been possible to accommodate a passenger in the open cockpit with some minor rearrangement. The aircraft's gross weight was 650lb (295kg) and propulsion was provided by a 41hp (30.5kW) Church Marathon J-3, 4-cylinder piston engine driving a two-blade wooden propeller.

Having completed the aircraft and put fuel in the tank, Johnson made the mistake of trying to start the engine within the confines of the garage. What possessed him to do this is a mystery, but to make matters worse, the throttle jammed wide open and the aircraft began to move. It finally pushed against a wall and he was able to cut the power, but not before some damage had been done. In autumn 1934, Johnson moved the Uniplane, issued with the registration NC13680, to Stinson Airport.

Testing began almost immediately and Johnson completed about two hundred taxiing runs and brief lift-offs. Eventually, he managed to get his Uniplane to fly and completed two separate circuits of the airfield. His first landing was fairly hard, but it seems that on the second occasion he managed to damage the aircraft quite badly. He walked away with just minor injuries but his plans to use the aircraft for local advertising purposes were over.

Repairs were made during the winter and then Johnson decided to make various alterations to the control surfaces and finally ditched the twin tail fin arrangement in favour of a single fin, which was how he originally designed the aircraft. After relocating to Harlem Airport, Johnson made a lengthy series of short take-offs and landings before attempting a circuit of the airfield. He then made six separate circuits, all culminating in hard landings and damage to the aircraft.

Right: **This drawing shows the initial concept for an aircraft design filed by Cloyd Snyder in late 1930 as a US patent. Many aspects of this concept would be utilised in the Arup glider that was completed in 1932.** *US Patent Office/Bill Rose*

It was now becoming clear that he lacked engine power and in a final act of desperation, Johnson managed to boost the performance slightly by using a higher-octane fuel. He finally completed two low-altitude flights, but after another serious accident in 1936, he decided to call it a day. The aircraft was sold to pay for his hangar facilities and may have been broken up for scrap.

Johnson then became involved with the World Federalist Party and ended up on trial in 1942 for his pacifist beliefs. After two years in jail, he moved to Sonoma, California, where he ran a gas station until he retired. He continued to promote the World Federalist Party through a publication called *Sphere* from his home at the DeAnza Moon Valley Mobile Home Park until his death in 2006.

ARUP INC

Dr Cloyd L. Snyder (1891-1971) was an American podiatrist (chiropodist) with a passion for aviation, living in South Bend, Indiana. In 1926, he observed – by accident – the way that a felt heel flew through the air and this led him to experiment with several small flying models, which were occasionally attached to the roof of his car for test purposes. On 8 September 1930, Dr Snyder filed a US patent for his promising aircraft design (1855695) and established a small aviation company called Arup Inc at South Bend.

Snyder created the name of his company by amalgamating the words 'air' and 'up' which described his aims fairly well. The country was entering the Great Depression

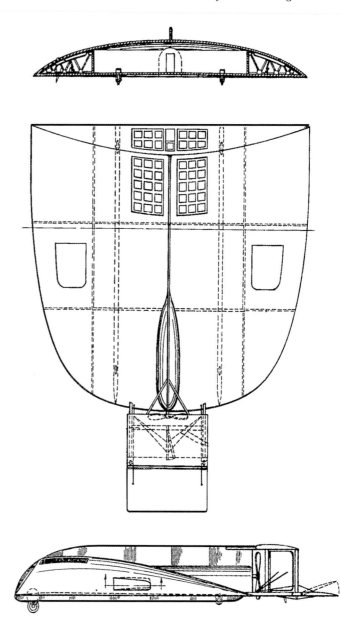

and money was tight, but Snyder intended to perfect an aircraft that was the aviation equivalent of the Ford Model T and then make the skies accessible to America's middle classes. If this proved successful, Snyder had further ambitions to build a long-range, heel-shaped transport aircraft with a wingspan of at least 100ft (30.48m) that would be 15ft (4.57m) thick at the highest point. Work was soon under way at Arup Inc to construct a small heel-shaped proof-of-concept demonstrator, which was a good deal smaller and more basic than the design he patented.

It was often referred to as the Dirigiplane as Snyder had considered the idea of installing bags filled with a lifting gas such as hydrogen. Although unusual in appearance, the unpowered, one-man Arup S-1 was built mainly from wood in a very simple manner. The wingspan was 19ft (5.8m), but no other dimensions are currently available. There was no windshield or canopy to protect the pilot, who was positioned in what passed for a cockpit at the centre forward section of the aircraft. The aircraft was fitted with two vertical stabilisers equipped with rudders and ailerons. In addition, there was a rear elevator section and the glider was fitted with a very simple fixed undercarriage, which probably consisted of two skids.

The S-1 was completed in early 1932 and gliding tests were then conducted at South Bend. About forty flights were made that proved the glider's stability and validated the STOL concept. It was then decided to modify the S-1 to accommodate a small Heath-Henderson internal combustion engine, although it is said that this addition was not a great success. Having hired an Austrian called Raoul Joseph Hoffman as the company's chief engineer, design work began on the Arup 2 (also called the Arup-Snyder 2), which was completed in 1933.

The Arup S-2 was more sophisticated in every respect compared to the glider, having a fully enclosed cockpit, better control surfaces and a fixed undercarriage with wheels. The S-2 was powered by a 36hp (26.8kW) Continental A-40 engine. With a gross weight of approximately 780lb (354kg), this allowed a maximum speed in

level flight of about 97mph (156kph). During trials, wingtip ailerons were added and repositioned on several occasions. The wingspan was 19ft (5.8m) including the 'ears' and the S-2's length was 17ft (5.18m). Utilising a NACA M-6 aerofoil, the S-2 enjoyed low take-off and landing speeds and was highly resistant to stall, although visibility from the cockpit could have been better. Snyder flew the aircraft on several occasions, as did James Doolittle, and it was displayed at the 1933 Chicago Air Races. The S-2 was also demonstrated to the military in Washington DC and loaned to NACA (National Advisory Committee on Aeronautics – the forerunner of NASA) who undertook limited testing at Langley.

Having assisted Snyder with design work on the S-3, Hoffman left Arup in 1934, moving to St Petersburg, Florida, where he built an Arup-style aircraft for the Chicago cosmetics millionaire James Leslie Youngblood. The Hoffman design incorporated several improvements over the S-2 and was constructed from welded steel and spruce. Propulsion was provided by an 85hp (63kW) Cirrus Mk 111 engine driving a wooden, two-blade 7ft (2.13m) propeller. The wingspan was 22ft 8in (6.9m), with an area of 237sq ft (22m²). The aircraft's length was 17ft 8in (5.38m) and the gross weight is recorded at 900lb (408kg). Hoffman planned to equip the aircraft with a fully retractable undercarriage, but this never happened, either for technical or financial reasons. The aircraft performed very well, with a speed range of 28-135mph (45-217kph) but during a flight in 1936, there was an engine fire and it crashed.

The Arup S-3, completed in 1934, was a marginally larger version of the Arup S-2, with a wingspan of 22ft (6.7m) and a length of 17ft 6in (5.33m). The ailerons were

positioned close to the wingtips and a 70hp (52kW) LeBlond 5DE engine was used to drive a two-blade wooden propeller.

Arup S-3 completed one test flight on 15 July 1934 at St Joseph County Airport before being destroyed in an unexplained fire that may have been arson. Snyder now began work on his next aircraft, Arup S-4, which was virtually another S-3 but with some minor improvements to the control surfaces. The S-4 flew for the first time on 19 March 1935. There were no major problems with the design and it performed without any significant problems until the start of World War 2.

Snyder had continued to design aircraft and his most advanced pre-war proposal was a low aspect ratio twin-engined monoplane, although it never progressed very far.

When America entered the war, Arup Inc was closed down. There are three different reports about the fate of S-2 and S-4. One suggestion is that both planes were auctioned and purchased by private buyers. It has also been said that vandals destroyed the aircraft. Both appear incorrect and it is likely that they were donated to a scrap metal drive to assist the war effort.

The basic Arup design was sound and in a re-run of history it is probable that these compact aircraft would have proved more successful. Nevertheless, the Arup design would make quite an impression on the aerodynamicist Charles Zimmerman, who would eventually conceive the Vought Flying Pancakes. Following the closure of Arup, Snyder joined the Goodyear Aircraft Corporation in Ohio. After his retirement, he moved to Plymouth, Indiana, where he died in 1971. As a footnote to the Arup story, a scale replica of Arup S-2 was completed in 1985 and successfully flown.

Right: **The first Arup aircraft, known as the S-1, was a glider constructed largely from wood and covered in fabric. After successfully completing about 40 test flights, it was equipped with a small Heath-Henderson internal combustion engine and two-blade propeller. The pilot seated in the aircraft remains unidentified, but the figure standing at the front of the aircraft has been identified as Cloyd Snyder.** *Bill Rose*

Left. **The completed Arup S-2, minus its wingtip ailerons, often referred to as 'ears'.** *Bill Rose*

Below: **The Arup S-2 at the Chicago Air Races in 1933.** *A. B. Bradley*

Two Arup aircraft parked outside the company's hangar at South Bend, Indiana. This picture is believed to have been taken in 1936. The aircraft on the left carrying advertising for Sears-Roebuck & Co is the Arup S-4 (14529). The second aircraft is the earlier Arup S-2 (12894). *Bill Rose*

Flying Saucer Technology

Above: **Forward view of the Arup S-2 as it is prepares for flight at the Chicago Air Races in 1933.** *A. B. Bradley*

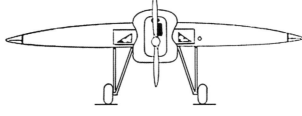

Right: **The design for a light aircraft produced by Raoul Hoffman after he left Arup, used to build a monoplane for J. Leslie Youngblood of Chicago in 1934. The aircraft was essentially an improved version of the Arup S-2.** *Bill Rose*

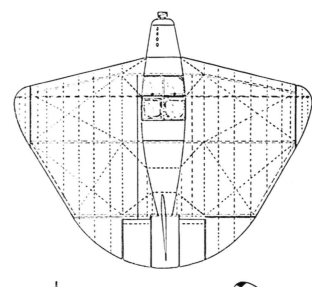

Below: **A poor quality photograph of Dr Cloyd Snyder (left) and Raoul Hoffman (right) during a public demonstration of the Arup S-2.** *Bill Rose*

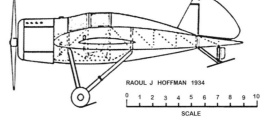

RAOUL J HOFFMAN 1934

0 1 2 3 4 5 6 7 8 9 10

SCALE

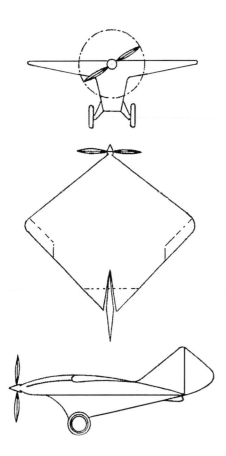

PIANA CANOVA DESIGN

The Arup aircraft also appears to have influenced the Italian aircraft designer Piana Canova, who filed patent No 457281 for an all-wing aircraft in 1935. Looking rather like a diamond-shaped version of an Arup design, it was wind tunnel tested as a scale model with good results.

At the time the patent was filed, Piana Canova claimed that a full-sized manned demonstrator was under construction. Powered by a single air-cooled Aeronca 35hp (26kW) piston engine, this very compact aircraft had a wingspan of 16ft 5in (5m) and measured exactly the same in length. The wing area was quoted at 141sq ft (13m²) and maximum speed was estimated at about 110mph (177kph), with cruise at 93mph (150kph) and a range of 650 miles (1,046km). The aircraft used a fixed undercarriage and the cockpit was located just behind the engine. A single tail fin with rudder was fitted and there were control surfaces towards the trailing edge wingtips. It is possible that the aircraft was completed and test-flown before the start of World War 2, but no documentation has been found to support this.

PAYEN AP.10/12

Another design with an affinity to Arup aircraft was produced in 1934 by the French engineer Nicolas Roland Payen (1914-2004). Payen created a very compact single-seat aircraft called the AP.10.

The AP.10 first flew at Dieppe in 1935, powered by an AVA 25hp (18.6kW) engine. The diminutive AP.10 had a wingspan of 16ft 3in (4.95m) and a length of 13ft 8in (4.16m). Wing area was 107.6sq ft (10m2) and when empty, AP.10 weighed about 441lb (200kg). The undercarriage was fixed and the aircraft utilised a single tail fin and rudder. It is believed that Payen replaced the piston engine with a more powerful unit in late 1935 and the AP.10 was flown throughout 1936. Handling was apparently good, with a modest maximum speed and good STOL performance. Payen envisaged a larger two-seat variant with an increased span of 19ft (5.8m), using the more powerful engine. He called this proposal the AP.12, but it never progressed beyond the design stage and Payen abandoned work on AP.12 in the late 1930s.

FARMAN F.1020

Henri Farman (1874-1958) and his younger brother Maurice (1878-1964) established a successful aircraft manufacturing business at Billancourt, Brittany, in 1908. As inventors, they were responsible for many engineering innovations and their company produced about two hundred different types of aircraft. The brothers were also motor racing enthusiasts who were briefly involved with the limited manufacture of prestigious sports cars. This part of their business came to an end in 1930, due to the global recession.

Despite this, the aviation company continued to make money and during the early 1930s, the Farman Brothers began to look at new designs. One experimental model was a light aircraft which appears to have been heavily influenced by the American Arup designs.

Design work on the Farman F.1020 probably started in early 1933 and construction was completed by the start of winter. The registration F-AMOG was issued

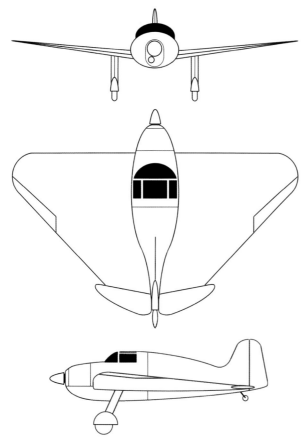

Above left: **Designed by Piana Canova, this small monoplane was designed in the mid-1930s and appears to have been influenced by earlier developments in the United States.** Bill Rose

Left: **The Payen AP.12 was a pre-war proposal based on the earlier AP.10. Larger and more sophisticated, it never reached the construction stage.** Bill Rose

for the aircraft and test flights commenced in December 1933. Described as a semicircular high-wing monoplane, the fuselage of this aircraft seems to have been very similar to the Farman F.400 series, which was a three-seat utility aircraft, developed at the same time. Propulsion for the F.1020 was a Lorraine-Dietrich 5Pb, 5-cylinder radial engine, providing a maximum of 125hp (93kW) at take-off. This engine was used to power the Farman F.402, and a wooden two-bladed propeller was also fitted. The F.1020's cockpit layout and flight controls are unknown, but it is possible that the control stick was suspended from the roof like the F.402.

The most interesting feature of the F.1020 is the semicircular wing fitted with substantial flaps and wingtip ailerons. Whether or not these were modified during testing is unknown. The tailplane arrangement looks identical to an F.400 series aircraft. Fuel was probably carried within the wing and the fixed undercarriage consisted of two forward wheels and a tail skid. The F.1020 is described in available literature as having two seats, although it is probable that enough room existed in the rear of the cockpit for a third person. The aircraft's overall wingspan was 23ft 7in (7.2m), with a length of 27ft 2in (8.27m). The maximum speed is thought to have been about 124mph (200kph), although minimum speed is unknown. Ceiling and range are thought to have been similar to the F.402.

It seems certain that the Farman F.1020 was

built to explore the option of producing an aircraft with superior STOL performance to the F.400 series. Very little is currently known about the development or testing of this aircraft, but it is thought that trials continued until the company was nationalised by the French government in 1936.

TRENN DUCTED FAN AIRCRAFT

Paul Trenn, a German engineer and inventor living in Wernigerode, specialised in the development of bicycles and motorcycles. In the late 1920s, he turned his attention to the idea of using two-stroke engines for driving the rotors of a small aerial platform.

Having refined the basic elements of his idea into a manned ducted fan VTOL flying vehicle, Trenn applied for a German patent and details of his invention were finally published on 25 October 1933 (Deutsches Reich Patent 586411).

This comprised a central upright cylindrical section surrounded by a substantial annular ring attached by cruciform bracing. Located within the area between the central section and the outer ring were four separate engines, each driving a rotor. Fuel for the engines would be carried in the annular ring. At a rough guess, the overall diameter of the aircraft would be about 16ft 5in (5m).

Shutters above or below the rotors were

Above: **The Farman F.1020 was devised in the early 1930s to provide improved STOL performance. Development of this interesting aircraft continued until 1936 when the Farman Company was nationalised.** *Bill Rose*

Below: **This rather unusual aerial platform was devised by inventor Paul Trenn at the end of the 1920s and published as a German patent on 25 October 1933. It remains unknown if Trenn built any models of this concept, although it seems unlikely that a full-sized aircraft was constructed.** *Deutsches Reich Patent Office*

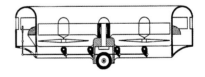

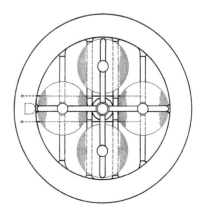

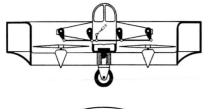

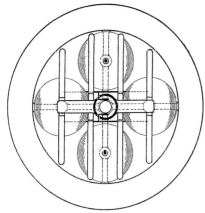

proposed to direct the airflow and provide flight control. The cockpit was located above the centre section in a dome-shaped enclosure or alternatively at the outer edge of the ring. The undercarriage was simple and rather crude, consisting of a single wheel with tyre directly below the central unit. At rest, the aircraft would tilt onto a section of the annular ring.

It is highly likely that Trenn experimented with models of his design before proceeding with a patent. However, the construction of a full-sized prototype would have presented many complex challenges. The vehicle would have required adequately

Left: **An alternative design for Trenn's manned ducted fan VTOL aircraft. This uses a slightly different system for mounting the rotors and a centrally positioned cockpit enclosure. It remains unclear exactly what use Trenn envisaged for this unusual design.** *Deutsches Reich Patent Office*

Below: **This early ducted fan VTOL aircraft was designed by Alex Andersen in 1930 and although the concept appears rather crude, it proved influential with future designers.** *US Patent Office*

powerful engines (probably of a radial design) and controlling the aircraft in flight would have been extremely difficult. It is unclear what purpose Trenn envisaged for this design and I have been unable to uncover any records of construction of a prototype. This cannot be entirely ruled out, but it seems unlikely that any manufacturer would have funded such an endeavour.

On 14 November 1930, Alex Andersen, an American designer in Seattle, Washington, filed a US patent for an aircraft with a single two-bladed propeller in a tiltable duct. Both the Trenn and Andersen designs would prove quite influential with future designers.

In the case of Trenn's flying machine, it was studied in some detail during the post-war years by engineers at companies such as Rolls-Royce and Glenn Martin. While interest generally focused on specific design features, the French aerodynamicist Jean-Marie Brocard continued to develop Trenn's concept, producing a series of proposals that bridged the gap between the original design and later unmanned aerial vehicles such as the Cypher.

FINK VTOL AIRCRAFT

Sharing some design similarities with Trenn's concept was an aircraft proposed by Abraham S. Fink of Aero Improvements Inc, New York. A US patent was applied for on 4 May 1935 (2077471) and was published in 1937. This was a twin-duct VTOL aircraft with helicopter capabilities and superior performance in level flight.

Two separate ducts provided propulsion, each containing an internal combustion engine and a twin-bladed rotor. Each rotor would turn in an opposite direction to its counterpart to counteract torque. In the initial proposal, Fink equipped the ducts with movable vanes for directional control. This would allow VTOL operation and directional horizontal flight. The rear of the craft was fitted with a single stabilising fin, although there is no mention of equipping it with a rudder.

There were several alternative layouts for this design. Fink considered the possibility of four separate ducted fans in two configurations (one almost circular) and the use of propellers for horizontal flight. The forward-positioned cockpit nacelle appears to have been designed for at least two occupants and would have provided excellent visibility. On the ground, the aircraft was supported by four fixed struts and wheels.

The intended role of Fink's aircraft is

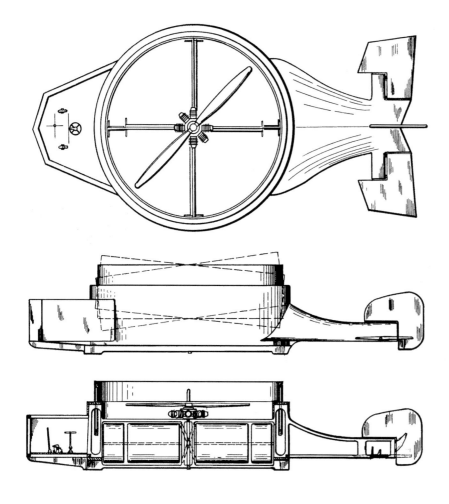

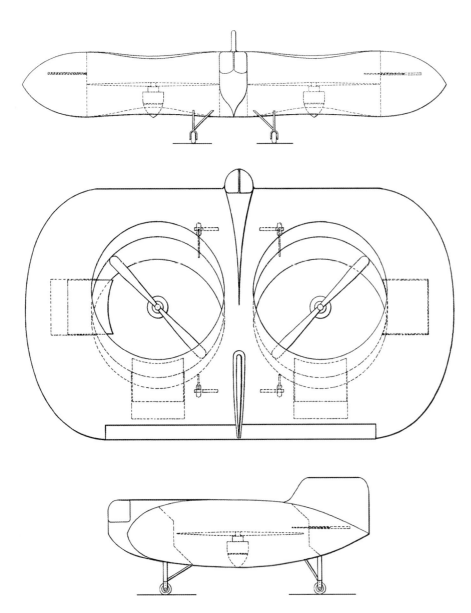

turbines developed by Frank Whittle and Hans von Ohain. Coanda called his system a Thermojet. It consisted of a steel panelled duct with its entrance in the nose of the aircraft. The duct enclosed a 4-cylinder 50hp (37kW) piston engine turning a compressor fan at 4000rpm. The airflow through the duct was controlled by a large adjustable iris (called an orburator) which was regulated by the pilot. Air and engine exhaust flowed along the duct and entered two ring-shaped combustion chambers where more fuel was added and burnt creating thrust. The exhaust gas then left the rear of this small aircraft.

On 16 December 1910, a public demonstration of the aircraft was undertaken at the Issy-les-Moulineaux Airfield in Paris. Coanda piloted the prototype and it made a spectacular start, gathering speed in a long trail of smoke. He began to climb into the air but suddenly lost control and crashed. Coanda escaped with minor injuries, but the aircraft caught fire and was destroyed. After this, there was little interest in his propulsion system and Coanda abandoned any further attempts to develop the Thermojet.

In the immediate aftermath of the accident, Coanda noticed the way that flames and incandescent gas tended to flow along the fuselage; many years later he would turn this chance observation into a series of experiments with jets of steam. Gases tended to flow around nearby curved

Below:**Henri Marie Coanda, one of the great aerodynamicists and inventors of the 20th century.**
Bill Rose

unclear as its payload and range would probably be quite limited. There is also the question of how the aircraft would have behaved in the event of engine failure. Not particularly aerodynamic, this concept lacked suitable propulsion technology and adequate methods of flight control. Needless to say, it progressed little further than the drawing board.

HENRI COANDA

Henri Marie Coanda (1886-1973) grew up in Bucharest, joined the School of Military Artillery, graduated as an artillery officer and developed an interest in aviation. He then studied at the Montefiore Institute in Liege, Belgium, and at the Superior Aeronautical School in Paris.

In 1910, Coanda built an aircraft in Joachim Caproni's workshop, which used a crude form of jet propulsion, although the design had little in common with later gas

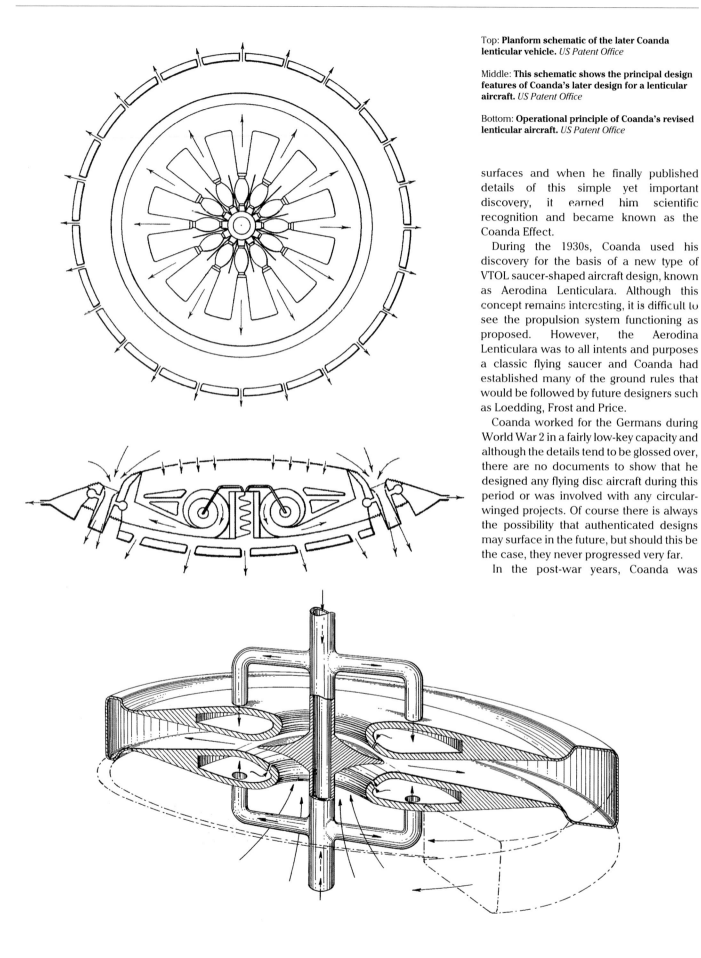

Top: **Planform schematic of the later Coanda lenticular vehicle.** *US Patent Office*

Middle: **This schematic shows the principal design features of Coanda's later design for a lenticular aircraft.** *US Patent Office*

Bottom: **Operational principle of Coanda's revised lenticular aircraft.** *US Patent Office*

surfaces and when he finally published details of this simple yet important discovery, it earned him scientific recognition and became known as the Coanda Effect.

During the 1930s, Coanda used his discovery for the basis of a new type of VTOL saucer-shaped aircraft design, known as Aerodina Lenticulara. Although this concept remains interesting, it is difficult to see the propulsion system functioning as proposed. However, the Aerodina Lenticulara was to all intents and purposes a classic flying saucer and Coanda had established many of the ground rules that would be followed by future designers such as Loedding, Frost and Price.

Coanda worked for the Germans during World War 2 in a fairly low-key capacity and although the details tend to be glossed over, there are no documents to show that he designed any flying disc aircraft during this period or was involved with any circular-winged projects. Of course there is always the possibility that authenticated designs may surface in the future, but should this be the case, they never progressed very far.

In the post-war years, Coanda was

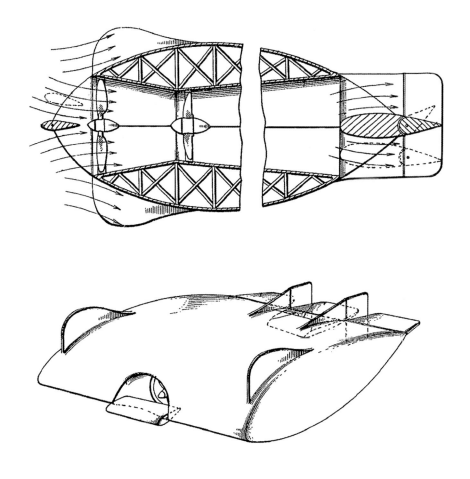

Left: **A relatively advanced monoplane design study produced by Dr Louis Crook in the early 1940s. A ducted propulsion system shown in the upper drawing was proposed, which might have eventually been replaced by a turbojet.** *US Patent Office*

advantages such as increased stability. He then turned his attention to propulsion, determining that the low aspect ratio airfoil was suited to thrust augmentation (which would now be described as an internal ducted fan arrangement) and perhaps the eventual use of one or more gas turbines.

By the end of the 1930s, Crook had completed a series of carefully considered designs evolving into a heel-shaped aircraft that employed ducted fan propulsion. It is clear that he understood the advantages of configuration that would allow the full utilisation of internal space, although he was more interested in establishing basic design features than producing a specific aircraft proposal.

His drawings show no provision for a cockpit, which might have needed offsetting due to the central duct and no undercarriage. The studies were completed in 1942 and, perhaps due to wartime restrictions, they appear to have been classified as secret until the end of hostilities. This interesting and advanced development work generated little interest until recently, but was certainly influential with aerodynamicists.

Filed Feb. 19, 1942

Louis H. Crook

CALDWELL DISC-ROTOR PLANE

In late August 1949, there were reports in the US press that wreckage from two man-made flying saucers had been found in an abandoned tobacco barn at Glen Burnie, near Baltimore, Maryland. Special agents from the USAF supported by local police officers were sent to examine this discovery, following enquiries that started several months earlier. They had been tipped off that a long-defunct company who developed experimental aircraft had once used the barn. This organisation had come to the USAF's attention after a former shareholder contacted them, claiming the company built flying saucer-type aircraft during the 1930s.

Following the well-publicised discovery at Glen Burnie, it appears that the USAF investigators were uncertain how to proceed. According to the *Baltimore Sun* newspaper, an unnamed USAF official said that the flying machines might have been early prototypes of flying saucers that were being sighted across the country. However, within twenty-four hours, the Air Force stated that this was not the case. They said an inventor had built these two experimental helicopter-type aircraft more than ten years earlier and there was no connection

invited to present a lecture on his flying saucer concept at Wright-Patterson AFB and he continued to develop his unusual design for a flying saucer, applying for a new French patent in early 1957. In 1970, he returned to Romania, becoming Director of the Institute for Scientific and Technical Creation, and he died in Bucharest on 25 November 1972.

CROOK'S DESIGN STUDIES

Although his name seems to have been largely forgotten, Professor Louis Henry Crook (1887-1952) was a truly brilliant inventor, designer and solver of complex engineering problems.

He worked as an aerodynamicist for the Naval Aerodynamical Laboratory, having expert knowledge of wind tunnel design and stress analysis. Crook also acted as a

scientific consultant for a number of different organisations and taught theoretical mathematics at the Catholic University of America. He wrote many papers on aviation issues that included pure aerodynamics, the development of guided missiles and supersonic flight. Professor Crook carried out detailed research into boundary layer control and was responsible for designing interference suppression measures for spark plugs and ignition systems. In June 1949, he won a legal action against the US government for patent infringement of this technology, worth millions of dollars.

During the 1930s, Professor Crook became interested in the design of Arup-type aircraft and this led him to examine the possibility of improving on the basic configuration. It was clear that the low aspect ratio design produced a low lift coefficient at small angles of attack, but he noted a number of useful

Above: **Discovered in a Maryland barn and described in some publications as the remains of a man-made flying saucer, the circular object seen in this photograph is actually the wing section of an unusual autogyro-type aircraft developed by Jonathan Caldwell in the late 1930s.**
US National Archives

to the recent wave of UFO sightings.

The inventor was Jonathan Edward Caldwell (1883-19??), a Canadian-born aeronautical engineer who had been responsible for producing several strange and completely unworkable concepts that were funded by very dubious methods. The elusive Caldwell would frequently move to different parts of the country, seemingly to avoid fraud charges, and in 1934, he arrived in Washington DC to start work on two new aircraft called the 'disc-rotor plane' and a revived version of an earlier design called the Cyclogyro.

After renting an office three blocks from the White House, Caldwell hired J. Owen Evans, who was an aeronautical engineer, and a skilled mechanic called Willard E. Driggers. Caldwell also retained the services of Dr Louis Crook who arranged wind tunnel testing of the Cyclogyro.

Construction of the disc-rotor plane was soon under way. This resembled an autogyro with a forward-mounted 9-cylinder radial engine driving a two-bladed propeller. The airframe was built from welded steel tubes and was mainly covered by fabric. The most unusual feature of this small single-seat aircraft was the overhead rotating circular wing. Its span is believed to have been 12ft (3.65m), with the hinged blades increasing the diameter to about 16ft (4.8m).

In forward flight, the airstream would cause the four short blades to spin the rotating wing, with fabric airfoils on the inside of the rim providing lift. Having reached an operational height, the wing would be braked for level flight. In theory, this would provide better performance than a helicopter, with an estimated maximum speed of about 100mph (160kph). On arrival at the destination, the wing would be unbraked, allowing an almost vertical landing.

The aircraft was controlled in flight by an unusual set of ailerons fitted to the tail section and a rudder. Initially, the fixed undercarriage comprised two forward struts with wheels and a simple tail skid. Caldwell's team completed the disc-rotor plane in 1937 and it was issued with a Civil Aeronautics Agency (CAA) registration number NX-99Y, indicating its experimental nature.

Left and below: **Another of Caldwell's rather bizarre VTOL experiments, sometimes referred to as the 'Cheesebox'. It was soon abandoned.**
US National Records, sourced by Joel Carpenter

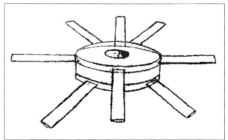

Flying Saucer Technology

It seems that Caldwell's shareholders were losing patience with delays and insisted on seeing a demonstration of the disc-rotor plane. In late 1937 (or possibly early 1938) Caldwell managed to talk Willard Driggers into making the first test flight. Driggers was a good engineer but had never flown any type of aircraft before, so it was a high-risk undertaking. The abandoned local Benning Racetrack was chosen for this demonstration and the aircraft was transported there and prepared for flight.

Apparently, Driggers opened the throttle and began to move forward at increasing speed. He was soon airborne, supported by the rotating wing and reached an altitude of about 40ft (12m). He then tried to manoeuvre, but found the flight controls were unresponsive. Fearful of gaining further altitude, Driggers panicked and throttled back, causing the aircraft to make a very hard landing about 600ft (182m) from the lift-off point. He walked away from the aircraft uninjured, but the undercarriage was badly damaged. This would be replaced by a tricycle arrangement, although the disc-rotor plane would never fly again.

Caldwell had also been developing another rather bizarre and unsuccessful aircraft, derived from one of his early VTOL designs. Now referred to as the Ornithopter, this aircraft consisted of two lengthy three-bladed aerofoils attached to the sides of a fairly conventional fuselage section. These were powered by a 125hp (93kW) radial engine. This prototype was abandoned, although Caldwell already had ideas for a new VTOL aircraft.

Having lost interest in the disc-rotor plane and run out of money, Caldwell relocated from Washington DC to Baltimore in 1939. Driggers and Evans were fired and Caldwell's 'Gray Goose Corporation' was closed. He now established a new company called Rotor Planes Inc, set up a workshop at Edmonson Avenue, Baltimore, and hired John W. Ganz, who was a local mechanic.

Caldwell's aim was to build a new VTOL aircraft which utilised two upper and lower counter-rotating rotors powered by an automobile engine. The prototype was completed in the barn at Glen Burnie and then secured in a framework. Several demonstrations followed and the aircraft is said to have hovered with some success. According to aviation historian Joel Carpenter, who has thoroughly researched the story, "this never could have flown. If it had, however, it would have looked saucer-like."

At some point in 1940, the Maryland Securities Commission began to take an interest in Caldwell and one day he simply disappeared with his family, leaving the two aircraft in the barn. Little was heard of him after this and it has not been possible to find an official record of his death in the US.

Below: **The very unusual experimental Caldwell disc rotor plane. The exact date of this photograph is unknown, but it was probably taken during 1937.** *Joel Carpenter*

Lee-Richards No 1 Circleplane

Accommodation: 2
Wingspan: 20ft (6m)
Length: N/A
Height: N/A
Powerplant: 7-cylinder 80hp (59.6kW) Gnome Lambda rotary engine
Gross weight: N/A
Maximum speed: 75mph (120kph) approx

McCormick-Romme Umbrellaplane

Accommodation: 1
Wingspan: 35ft (10.6m) approx
Powerplant: 7-cylinder 50hp (37.28kW) Gnome Omega rotary internal combustion engine
Maximum speed: 50mph (80kph) estimate
Range: N/A

Nemeth Roundwing

Accommodation: 2
Wingspan: 16ft (4.8m)
Length: 20ft (6m)
Height: N/A
Powerplant: 90hp (67kW) Lambert engine, replaced by 7-cylinder 125hp (93.2kW) Warner Scarab 50 engine
Landing speed: 30mph (48kph)
Maximum speed: 140mph (225kph)
Ceiling: N/A
Range: N/A

Uniplane

Accommodation: 1
Length: 16ft 6in (5m)
Wingspan: 14ft (4.26m)
Powerplant: 41hp (30.5kW) Church Marathon J-3, 4-cylinder piston engine
Gross weight: 650lb (295kg)

Arup S-2

Accommodation: 1
Wingspan: 19ft (5.8m) including the 'ears'
Length: 17ft (5.18m)
Height: N/A
Powerplant: 36hp (26.8kW) Continental A-40 engine
Empty weight: 489lb (222kg)
Gross weight: 780lb (354kg) approx
Maximum speed (level flight): 97mph (156kph) approx
Range: N/A

Arup S-4

Accommodation: 2
Wingspan: 22ft (6.7m)
Length: 18ft 6in (5.63m)
Powerplant: 70hp (52kW) LeBlond 5DE engine
Empty weight: N/A
Gross weight: N/A
Maximum speed: 120mph (193kph)
Minimum speed: 30mph (48kph)
Ceiling: 9,000ft (2,743m)

Canova

Accommodation: 1
Wingspan: 16ft 5in (5m)
Wing area: 141sq ft (13m²)
Length: 16ft 5in (5m)
Powerplant: Air-cooled Aeronca 35hp (26kW) piston engine
Maximum speed: 110mph (177kph)
Range: 650 miles (1,046km)

Payen AP-10

Accommodation: 1
Wingspan: 16ft 3in (4.95m)
Wing area: 107.6sq ft (10m²)
Length: 13ft 8in (4.16m)
Powerplant: AVA 25hp (18.6kW) engine
Empty weight: 441lb (200kg) approx
Gross weight: 750lb (340kg)
Minimum speed: 28mph (45kph)
Maximum speed: 124mph (200kph)

Farman F.1020

Accommodation: 2 (possibly 3)
Wingspan: 23ft 7in (7.2m)
Wing area: 290.62sq ft (27.0m²)
Length: 27ft 2in (8.27m)
Height: 7ft 3in (2.22m)
Powerplant: Lorraine-Dietrich 5Pb, 5-cylinder radial engine, providing a maximum of 125hp (93kW)
Gross weight: N/A
Maximum speed: 124mph (200kph)
Ceiling: 12,000ft (3,657m) approx
Range: 621 miles (1,000km) approx

Caldwell Disc-Rotor Plane

Accommodation: 1
Wingspan: 12ft (3.65m), increasing to 16ft (4.8m) with blades
Length: N/A
Powerplant: 9-cylinder radial engine
Estimated maximum speed: 100mph (160kph) in level flight

Wartime Developments

Towards the end of the 1930s, steadily improving aircraft engines were suggesting the near-future prospect of a compact circular-shaped aircraft with excellent STOL or even full VTOL performance. This had already attracted interest from the US Navy and Germany's Luftwaffe, which were both funding small-scale development projects.

FLYING PANCAKES

In the late 1920s, Charles Horton Zimmerman (1907-1996) was accepted as a student by the University of Kansas. Having chosen to study electrical engineering and basic aeronautics, he was then invited to participate in a wind tunnel project at NACA's Langley Memorial Aeronautical Laboratory.

The agency was building a new open-throat wind tunnel and Zimmerman's contribution to this project would lead to a full-time job in 1929. Having completed his work on this project, Zimmerman became involved in the development of two significantly larger wind tunnels, which were fully funded in 1933 and became operational in 1935.

Zimmerman was equally interested in aircraft design and took a close look at the Arup S-2, when it was tested by NACA

Langley in 1933. He believed it would be possible to build a twin-engined, elliptical or near-circular aircraft with exceptional performance in normal flight and the ability to lift off and land like a helicopter. Zimmerman developed his ideas into a firm proposal, which won him an award from NACA, although it was somewhat over-ambitious and many years ahead of what was technically achievable.

Nevertheless, Zimmerman was convinced that his idea could be made to work and on 30 April 1935, he filed a US patent (2108093) for a semicircular, low aspect ratio aircraft driven by two propellers.

Engine performance and weights are unknown, but Zimmerman anticipated a performance range of 0-200mph (0-322kph). The undercarriage was to be fully retractable and in addition to the centrally located prone pilot there would be accommodation for two passengers, also in prone positions. He then decided to build a small aircraft based on this design, and was assisted by Richard Noyes and John McKeller.

This aircraft was constructed mainly from wood. It had a 7ft (2.13m) wingspan and was powered by two 25hp (18.6kW) Cleone engines. Zimmerman designed this machine to be flown by a prone pilot, but it never happened due to engine synchronisation problems. Perhaps this was the best outcome, as the design was considered rather dangerous.

In 1936, Zimmerman was able to prove his concept, using an elementary 20in (508mm) rubber band-powered model. These experiments made quite an impression on United Aircraft Corporation's President Eugene E. Wilson, who offered Zimmerman a designer's job at the company's Chance Vought Aircraft Division, located at Stratford, Connecticut.

Although Zimmerman was now in charge of NACA's Stability and Control Section, this

offer was too good to turn down and he joined Chance Vought in late 1937. It was agreed that he would be allowed to pursue his own line of research, and Zimmerman began by constructing a model-sized, circular-winged aircraft that was designated V-162. The overall appearance of this model is recognisable as a Flying Pancake, minus certain features found on the later full-sized aircraft, such as the 'ailevator' fins attached to the wing's trailing edge. The overall span of the model is reported to have been 3ft (0.91m).

The V-162 was fitted with two upright stabilising fins, a fixed undercarriage and there was a slight central forward protrusion that represented a cockpit. Two propellers were linked by small drive shafts to two electric motors positioned inside the centre of the model and power was provided from an external supply. The angle of the hinged tail section could be adjusted by remote control and several cables tethered the model during trials. V-162 was successfully flown inside one of the company's hangars, demonstrating the potential for further development.

February 1938 saw the publication of Zimmerman's US patent (2108093) for a low aspect ratio aircraft, which he called the Aeromobile. Vought proposed development of this aircraft for military observation, issuing the designation V-170. The US Army were approached for sponsorship in October 1938 and they are reported to have been impressed with the concept, but turned it down.

Vought hastily attempted to generate commercial interest in a civil utility version known as the V-171 and finally returned to the US Army with an attack version called V-172. This also met with rejection, perhaps because Chance Vought was a major US Navy contractor and the US Army preferred not to cross any established boundaries.

Following internal changes and a relocation from East Hartford to Stratford, Connecticut, in early 1939, the company became known as the Vought-Sikorsky Aircraft Division. This would work to Zimmerman's advantage, as there were

Left: **An original photograph showing the NACA Langley 15ft Spin Tunnel building during construction in 1934. This is one of the projects that Charles Zimmerman worked on and it took him into the world of aircraft design.** *NASA*

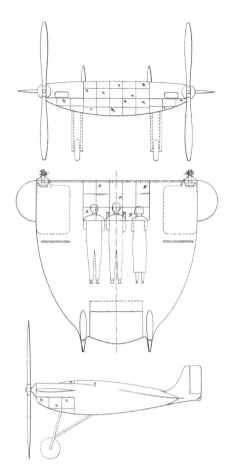

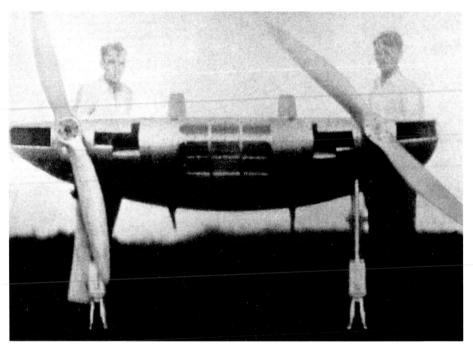

Above: **This is the original design for a semicircular, low aspect ratio, twin-engined utility aircraft proposed by Charles Zimmerman in the 1930s. Given the name Aeromobile, many features would be utilised in the later V-173 Flying Pancake.** *US Patent Office*

Above right: **This small, one-man aircraft was completed in the mid-1930s by Charles Zimmerman and several of his colleagues. Closely based on Zimmerman's Aeromobile design, there were serious engine balancing problems, so it never flew. This was probably for the best as the model was considered rather dangerous.** *Bill Rose*

Right: **The V-162 was a proof-of-concept scale model demonstrator built and flown by Charles Zimmerman at Vought's factory in 1937.** *Vought Heritage*

renewed efforts to sell his experimental design – this time to the US Navy.

Vought submitted plans to the Navy on 7 March 1939 for a prototype with the new company reference V-173 and, six months later, a scale model of the proposal was sent to NACA Langley for wind tunnel tests. Some aerodynamic problems would become apparent and part of the solution was to fit stabilising fins with control surfaces called 'ailevators' (aileron-elevator) to the trailing edge of the aircraft. The ailevators were initially tested on a small electrically powered model, which was similar in size to the V-162 and may have been a rebuild of this demonstrator.

Vought V-173

During spring 1940, the US Navy approved construction of the V-173 prototype and a contract was issued on 4 May.

In its initial form the V-173 was designed to be flown by the pilot lying in a prone position. A wooden mock-up of the cockpit was built, but it did not meet with widespread approval. So a redesign followed, which allowed the pilot to sit in an upright position with entry via a trapdoor in the floor. On 30 July 1941, the Navy issued serial number 02978 for the

completed V-173 and in mid-September 1941, it was shipped to NACA Langley for tests in the full-sized wind tunnel. Ernest Joseph 'Woody' Greenwood (1913-2007) had now been assigned to Zimmerman's project as chief engineer and he was instrumental in replacing the two-blade propellers with three-blade designs.

The V-173 was returned to Stratford in 1942, with engine testing and taxiing runs under way by the autumn. After a few near lift-offs it was decided that the aircraft, which

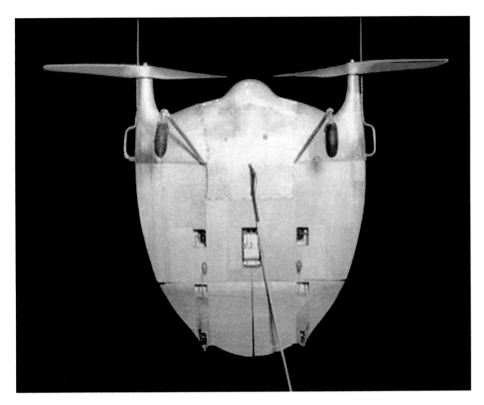

was now being referred to as 'The Flying Pancake', was ready for its first test flight.

It was constructed largely from wood and aluminium with a bright yellow colour being applied to the fabric covering. The wingspan was 23ft 4in (7.11m), with a wing area of 427sq ft (39.66m²) and an overall length of 26ft 8in (8.12m). Propulsion was provided by two 80hp (59.6kW) Continental A-80 piston engines driving 16ft 6in (5m) diameter propellers. Only 16.5 UK gallons (75 litres) of fuel was carried, but this was more than adequate for any test flight. While the V-173 was underpowered, its low gross weight of 2,258lb (1,024kg) and high lift was expected to allow take-offs in less than 200ft (60m), providing wind conditions permitting. However, it was necessary to raise the angle of the fixed position undercarriage to just over 22°, which interfered with aspects of the pilot's visibility and required the use of some forward glazing. On 23 November 1942, Vought's senior test pilot, Boone Guyton (1913-1996), successfully lifted off and made a circuit around the company's airfield. Accompanied by a chase plane, the V-173 completed its flight in 13 minutes. Although it performed reasonably well, Guyton reported that handling was poor and this was quickly traced to improperly balanced control surfaces. However, the aircraft's impressive

STOL performance had become evident when it landed on the runway in less than 50ft (15m).

Piloting the V-173 was unlike any other aircraft. With both engines running at full power and the stick pulled right back, it was possible to fly with a nose-up attitude of 45° while maintaining complete lateral and longitudinal control. It never stalled or showed any tendency to spin.

The maximum speed was about 138mph (222kph) and V-173 could cruise at 75mph

(120kph). The ceiling was approximately 5,000ft (1,524m) and range was limited, but this aircraft was never intended to set performance records. Nevertheless, this inadequate engine power created a tendency to fly nose-up, with the situation being reversed during landing. Consequently, all pilots who flew the V-173 were agreed that controlling the aircraft could be quite demanding.

Engine noise and resonance generated by the propellers made V-173 quite unpleasant to fly. It was possible to counteract the problem with vibration dampers, but the noise level remained high and pilots simply adjusted to it. In the course of testing this aircraft, there were two crash landings, although the damage was negligible in both cases.

The first incident occurred on 3 June 1943 when test-pilot Richard Burroughs experienced fuel vapour lock problems during an otherwise uneventful flight. His only option was to make an emergency landing on Lordship Beach, just below Bridgeport on Long Island Sound. As Burroughs touched down, narrowly avoiding a startled sunbather, the aircraft flipped over when its wheels sank into the soft sand. Burroughs was uninjured and managed to squeeze out of the cockpit after pushing some sand aside. Two propeller blades were broken, but there was little damage and V-173 was easily repaired.

The second accident took place on 26 May

Right: **The Vought V-173 Flying Pancake undergoes wind tunnel testing at NACA Langley in 1941.** *NASA*

Above: **Taken from an accompanying chase plane, the Vought V-173 is seen during its first brief test flight on 23 November 1942.** *US Navy*

Left: **Vought V-173 parked at the company airfield.** *US Navy*

Bottom: **The Vought V-173 during an early test flight.** *US Navy*

1945 when engine problems forced Burroughs to make an emergency landing on the Mill River Golf Course near Stratford. Again, the aircraft was relatively undamaged and it was covered over and placed under guard while awaiting collection. V-173 remained a classified wartime project, so it was finally transported back to the airfield under cover of darkness.

A third catastrophic accident was narrowly averted during Chance Vought's Thirteenth Anniversary Air Show in 1947. On this occasion, Guyton was taking off when he experienced unexpected lift problems and just managed to avoid several high-tension cables and a rock face at the end of the runway.

The V-173 completed a total of 131 hours in the air and was flown by Boone Guyton, Richard Burroughs, Charles A. Lindbergh and several US Navy pilots. Following its 171st flight in 1947, the project was formally ended. V-173 was eventually passed to the Smithsonian National Air and Space Museum, which stored the aircraft at its

Above: **A rare colour image of the Vought V-173 during an early test flight.** *US Navy*

Right: **On 3 June 1943, the V-173 made an emergency landing at Lordship Beach, Connecticut, following a fuel supply problem. Test pilot Richard Burroughs escaped without injury and although the aircraft flipped over in the soft sand, it was relatively undamaged.** *Bill Rose*

facility in Suitland, Maryland, until the end of the 20th century.

In 2000, the museum agreed that a group of former Vought employees belonging to the company's Heritage foundation could borrow the V-173 and fully restore it. Subsequently, the aircraft was partly dismantled, crated in a custom-built plywood and metal container (at a 32° angle) and shipped by flatbed truck across six states to Vought's facility in Dallas, Texas. The aircraft arrived in late 2003 and a team of 22 former Vought engineers (now in their eighties) began restoration work on the historic aircraft.

To make V-173 easier to work on, a substantial rotating steel framework was constructed. Now the aircraft could be turned to any position required. Both engines were fully overhauled, the cockpit Plexiglas was replaced and many small components were repaired or overhauled. Finally, the cotton fabric skin was replaced and the aircraft was repainted in its original yellow and silver scheme. The immaculate V-173 was

Above: On 3 June 1943, the V-173 made an emergency landing at Lordship Beach, Connecticut, following a fuel supply problem. Test pilot Richard Burroughs escaped without injury and although the aircraft flipped over in the soft sand, it was relatively undamaged. *US Navy*

Below: The V-173, showing revised markings and fitted with wheel spats, is prepared for take-off. *US Navy*

Bottom: V-173 prior to a test flight. Date unknown. *US Navy*

eventually exhibited at the Dallas Frontiers of Flight Museum, V-173 before its return to Washington DC.

Vought XF5U-1

The V-173 Flying Pancake was a proof-of-concept demonstrator that would be used as the first stage in a US Navy-sponsored project to develop an advanced STOL shipboard fighter. Charles Zimmerman started design work on a successor to the V-173 as far back as 1939 and plans were filed as a US patent on 18 December 1940 (2431293), which would remain classified until 18 November 1947.

This design steadily evolved into the Vought-Sikorsky Model 315 (VS-315), which was similar in size to the V-173 and slightly more streamlined. VS-315 would be fitted with more powerful engines and it was initially planned to utilise a prone position for the centrally located pilot.

The US Navy had informally accepted the VS-315 for further development and following a review they issued a Letter of Intent to Vought on 17 September 1942. This would allow work to begin on the construction of a mock-up, followed by two prototypes. One would be flown and the other was for destructive static testing.

This new aircraft would be officially known as the XF5U-1 and one of the first design alterations requested by the Navy was a conventional cockpit layout, with the pilot seated in an upright position. Like the V-173, control surfaces were located at the trailing edge and each vertical fin was fitted with a rudder.

The XF5U-1 would be the prototype of a carrier-based, long-range fighter-bomber. It was expected to have outstanding STOL performance and a future propulsion upgrade might allow it to hover in a nose-up position for brief periods. It was hoped that this ability would make the aircraft suitable for carriage on relatively small carriers and by some modified battleships. It was planned to link the engines to three-blade propellers, and these were fitted to the wooden mock-up first shown to a US Navy delegation on 8 June 1943 at Stratford.

However, it was becoming apparent that these propellers were completely unsuitable for use at high angles of attack. Unacceptable vibration would be created by asymmetrical airflow and the project was now in serious difficulty. By November 1943, Zimmerman had come up with a new propeller unit design. This would use articulated blades that would 'flap' about the axis of the drive shaft in pairs.

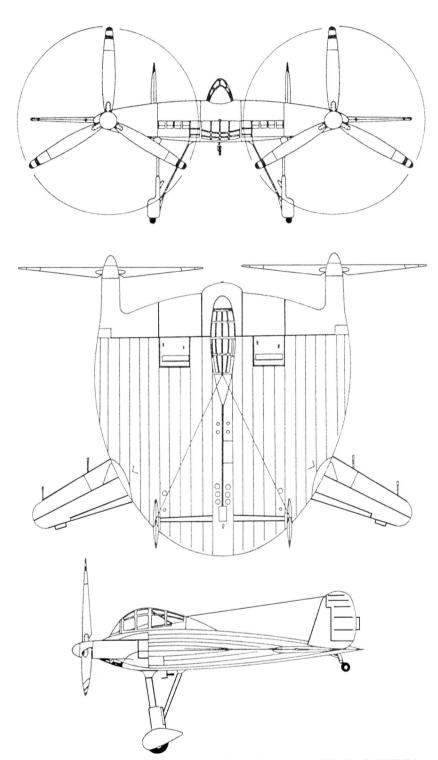

General appearance of the Vought V-173 Flying Pancake. *Bill Rose*

Above: **Vought's senior test pilot, Boone Guyton, was heavily involved with the testing and development of Zimmerman's V-173 and later XF5U-1 aircraft.** *Bill Rose*

linked to the propellers via a complex gearbox and drive shaft arrangement. In 1944, Zimmerman attempted to simplify the design with angled engines connected to gearboxes behind the propellers, but this idea was never taken any further despite showing promise.

Calculations indicated that the R2000-7 engines would provide a performance range of about 50mph to 425mph (80-683kph). Once the initial flight trials had been completed, it was intended to upgrade to more powerful Pratt & Whitney R-2000-2 (D) Wasp turbo-supercharged 14-cylinder units rated at 1,600hp (1,193kW). This was expected to allow landing speeds of 40mph (64kph) and a speed approaching 500mph (800kph) at an altitude of 30,000ft (9,144m), which was close to the theoretical limit of propeller-driven aircraft performance.

The XF5U-1 would be constructed from steel, aluminium and a newly developed composite material called Metalite, which was formed from balsa wood sandwiched between aluminium sheets. When completed, the aircraft had an empty weight of 13,107lb (5,945kg) and a gross weight of 16,802lb (7,621kg). When final calculations for the XF5U-1's performance were completed, the aircraft's 217 UK gallon (986.5 litres) fuel capacity was expected to provide a maximum range of about 900 miles (1,448km). This could be extended by the carriage of two 124 UK gallon (564 litre) drop tanks.

The XF5U-1 had a span of 32ft (9.75m) and an overall length of 28.6ft (8.71m). Its wing

Although these new 16ft (4.87m) diameter propellers were expected to resolve the problem, they would need to be custom made at Stratford. So a decision was taken to fit Hamilton Standard Hydromatic four-blade propellers as an interim measure. These would allow ground testing and taxiing trials to take place while the new units were being produced.

The initial choice for propulsion was two Pratt & Whitney R2000-7 radial air-cooled engines, rated at 1,350hp (1,006kW) and

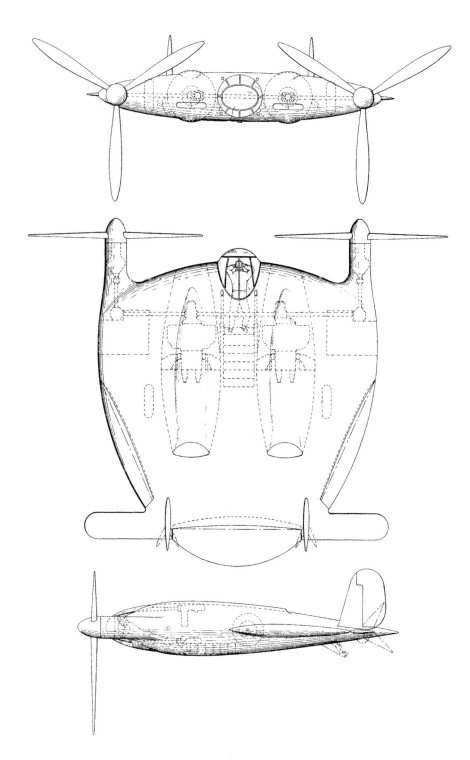

Above: **Charles Zimmerman.** *US Navy*

Left: **The Vought VS-315 was an intermediate unbuilt design that bridged the V-173 and XF5U-1. The prone cockpit layout was rejected by the US Navy.** *US Patent Office/Bill Rose*

being informally called the 'Zimmer Skimmer', sometimes the (second) 'Flying Pancake', or the 'Flying Flapjack'.

In August 1945, the XF5U-1 was rolled out and taxiing trials began at the company's airfield. Nevertheless, the complex engine and gearbox layout continued to cause headaches and these, combined with what Vought described as accountancy problems, are said to have caused further delays in the test schedule.

The four-bladed propellers were finally replaced by the new articulated units and Vought's helicopter specialist Ralph Lightfoot suggested fitting extended fairings on the outer sides of the propeller nacelles to improve aerodynamic performance, but this option was never taken up. The static XF5U-1 had been destroyed during tests to determine load limits and the flightworthy XF5U-1 was scheduled to begin flight-testing in 1947, with preparations under way to transport the aircraft to Muroc Field in California (now Edwards AFB) via the Panama Canal.

The choice of Muroc Field remains puzzling, as the obvious place to conduct trials was the Patuxent River Naval Air Station in Maryland. Arguments in favour of using Muroc suggest improved security and greater landing options, but neither of these appears to be a sensible reason for such an expensive and inconvenient relocation. That aside, the writing had been on the wall for some time and the Navy was fully aware that propeller-

area was exactly the same as the V-173 at 427sq ft (39.66m²) and the airfoil section chosen was NACA 0018. The aircraft was equipped with a retractable undercarriage that was angled upwards by 18° to provide improved lift during take-off. Although this was somewhat less than the V-173, there were still visibility problems during take-off and landing.

It was initially intended to equip the production fighter with six .50 cal (12.7mm) machine guns, which were later replaced by four 20mm cannons. These would be housed alongside the cockpit. There was also provision to carry two 1,000lb (453kg) bombs or additional fuel in drop tanks. The XF5U-1 flight prototype was issued with the US Navy registration number 33958 and it was now

driven combat aircraft were approaching obsolescence and could not compete with jets. This was also obvious to Vought, who were working on designs for gas turbine powered fighters.

In reality, continuation of the XF5U-1 project had been in doubt for some time and by early 1947, a decision had been made by the Navy to scrap the project. The contract was officially cancelled on 17 March 1947, bringing flights of the V-173 to an end and halting any further ground tests of the XF5U-1. Vought were relocating to a new facility in Dallas and company officials did not appear too concerned about the loss of what had been an expensive and troublesome project. The V-173 was eventually moved into storage and then passed to the Smithsonian, but the Navy was insistent that the XF5U-1 should be totally destroyed.

Having removed the engines, the aircraft was towed out of the hangar and passed to a wrecking crew. Chance Vought's senior test pilot, Boone Guyton, had strongly protested to the company's senior management and finally attempted to prevent the wrecking crew from carrying out their assignment, but it was all to no avail.

The work began and the XF5U-1 resisted the wrecking ball for some time, but it was finally smashed to pieces and ended up as part of a huge 40ft (12m) high pile of metal scrap that included jigs and special equipment used to build the aircraft. It was then realised that $6,000 worth of silver used for parts in the gearboxes ($6,000 was a good deal of money in 1947) was somewhere within the wreckage.

Staff and security personnel were immediately assigned to look for the silver, but it proved to be an impossible task and the search was finally abandoned. Nevertheless, the scrap dealer who hauled away the wreckage finally located the silver and this came to the FBI's attention when he tried to sell it. As a consequence, Vought were forced to compensate the Navy for loss of the silver and confirm the scrap dealer's legal right to profit from their mistake.

As a footnote to this story, there have been a number of claims by conspiracy theorists that the XF5U-1 was secretly shipped to

Top: **Vought XF5U-1. The second test example is visible in the rear of this picture.** *US Navy*

Middle:**The Vought XF5U-1 fitted with standard Hydromatic four-blade propellers, prior to initial taxiing trials.** *US Navy*

Bottom: **A rear view of the Vought XF5U-1 prototype, equipped with the articulated propellers.** *US Navy*

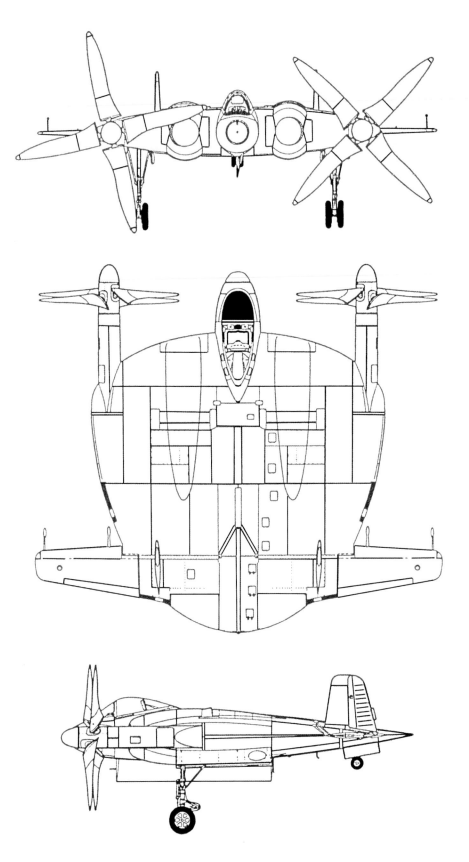

The advanced Vought XF5U-1 prototype shipboard fighter, fitted with the later articulated propeller blades. *Bill Rose*

Muroc field via the Panama Canal and it crashed during a test flight. There is absolutely no evidence to support this theory and it should be dismissed.

When Vought relocated to Texas in 1947, Zimmerman left the company and rejoined NACA at Langley where he resumed his work on wind tunnel testing. In 1952, he joined the Brown Hypersonic Group, set up by NACA to study proposals for trans-atmospheric vehicles, which finally led to the North American X-15 rocketplane. Six years later Zimmerman joined NACA's Space Task Group and in 1962 he became NASA's Director of Aeronautics, remaining in this post until 1967 when he retired. Zimmerman died on 5 May 1996.

ADVANCED PANCAKES

Author's note: Although the information below is completely at odds with historical records, I spoke and corresponded with Mr Smith in considerable detail, finding him a well-educated, honest and likeable person. I make no claims about the following story and ask readers to reach their own conclusions.

Having previously dismissed the idea that the XF5U-1 avoided destruction and reached flight-testing, there is the slim possibility that a highly classified parallel project existed, involving at least one other advanced version of Zimmerman's Skimmer. Thomas Clair Smith (1925-2005) graduated from Penn State University in 1946, having received a degree in mechanical engineering. He then secured a job with Vought at Stratford as an engineer who was responsible for testing various aircraft components.

About a month after joining Vought, Smith was vetted by the FBI and cleared to work on classified projects. He was then moved to a department, which was testing the characteristics of newly developed composites like Metalite in a pressurised high-altitude chamber. Apparently, security at the plant was extremely tight and work on new projects was often compartmentalised. However, Smith expressed an interest in the ultimate use of the material he was testing and because of his FBI clearance he was allowed to see several prototype aircraft in development.

The first aircraft he recalled being shown was a partly completed, classified jet fighter and the second was a VTOL version of the Skimmer that remains unacknowledged. Smith described this aircraft as semi-elliptical, fitted with two sets of propellers, two upright tail fins and standing on "stilt-type landing gear with wheels". He went on to say

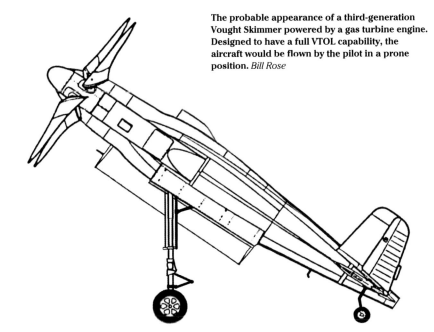

The probable appearance of a third-generation Vought Skimmer powered by a gas turbine engine. Designed to have a full VTOL capability, the aircraft would be flown by the pilot in a prone position. *Bill Rose*

the aircraft was piloted in a prone position and there was a transparent dome at the forward centre of the aircraft. Smith was convinced that the aircraft used gas turbine propulsion to drive the propellers and he thought he had been shown a separate development of the XF5U-1.

"Did the aircraft fly? You bet! I saw it take off, hover and land. It was only tested at night. They'd get it off the ground and it would disappear into the darkness." According to Smith, stories began to circulate about unidentified objects seen in the night sky near the company's airfield. "Those of us that worked on the project," said Smith, "got a chuckle about these reports, since we knew what they had seen."

Smith left Vought in 1947 when the company moved to Texas and he believed the prototype was taken there. Enjoying a

long and distinguished career, Smith eventually became Vice President of the Woodstream Corporation. He was listed in *Who's Who in America*, 54th edition. No documents exist to support Smith's intriguing story and had it originated from another source, the chances are that it would not be in this book.

However, the description of this aircraft fits two possible designs considered for a third-generation Skimmer. The first was based on the XF5U-1 and used a centrally located axial flow gas turbine that was shaft-coupled to each propeller. The engine exhaust gases would be vented through a port in the tail, so two tail wheels were required rather than one.

The second proposal involved a substitution of the two air-cooled radial engines for gas turbines, which would be

This concept artwork shows a civil post-war turbojet-powered Skimmer, intended as a multi-role VTOL utility aircraft. *Bill Rose*

interconnected, and shaft-coupled to each propeller. This second option, designated VS-341, was briefly mentioned in a Naval Aviation Confidential Bulletin for October 1946, which suggested a theoretical performance capability of 0-500mph (0-885kph).

Both of the gas turbine powered proposals would have used fully retractable forward struts that provided a substantial maximum elevation of around 35°. These double-action stilted legs would allow VTOL when fully extended and a rolling horizontal take-off when partly retracted. Zimmerman devised a number of very complicated undercarriages for advanced versions of the Skimmer, with the most elaborate allowing the aircraft to sit almost upright thanks to a complex swinging tail arrangement.

It is understood that Vought proposed an advanced turboprop Skimmer with a full VTOL capability for a joint US Navy-USAAF design study called Project Hummingbird, which started in April 1947. Little documentation can be found for this unsuccessful concept, but it probably incorporated most features already discussed and may have been a two-seater.

In the early post-war years, Zimmerman tried hard to generate commercial interest in a civil version of his VTOL Turbo Skimmer that would serve as a light utility transport. This design accommodated a pilot and two passengers in a swivelling cockpit that would remain horizontal during take-off and transition to level flight.

Nobody can say with absolute certainty if this aircraft might have been successfully developed, but the reliability of the power transmission system would have been questionable and the behaviour of the articulated propellers was poorly understood. It is certainly hard to imagine this aircraft realistically competing with a cost-effective 1950s fixed wing two-seat utility aircraft such as the Cessna 150.

In 1954, a Norfolk, Virginia-based aeronautical designer called John J. Sullivan produced designs for a new aircraft based on Zimmerman's earlier work. Sullivan proposed a VTOL aircraft with a rotating cylindrical cockpit intended to move through 90° and maintain a stable position for the crew during all aspects of the flight.

Somewhat larger than the Flying Pancakes, this aircraft would use a complex expanding support system while standing on the ground. Various methods of propulsion were considered including propellers and jet engines in pods. This strange development would have represented a major engineering challenge to build and soon slipped into obscurity.

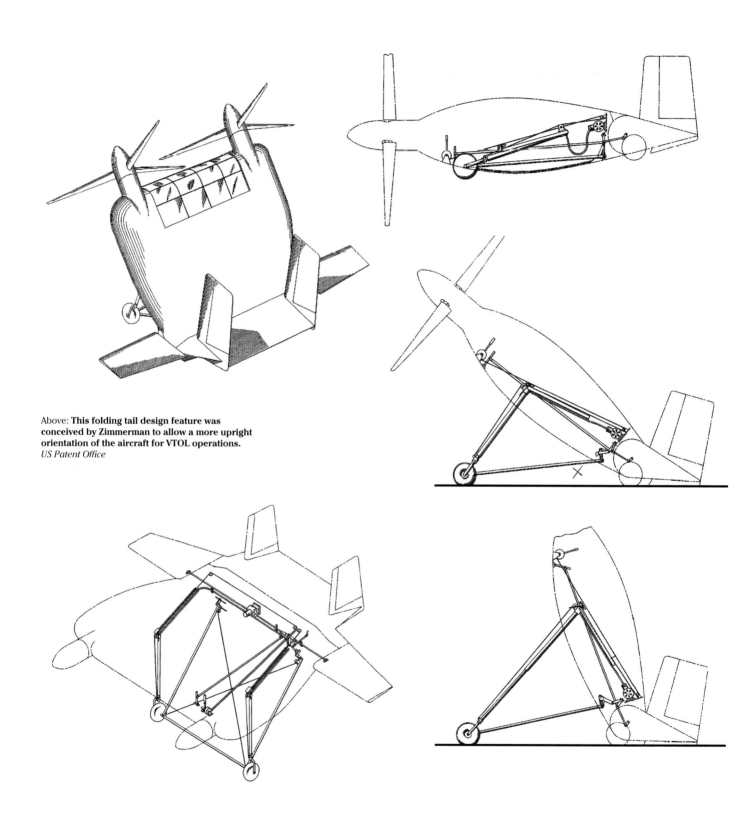

Above: **This folding tail design feature was conceived by Zimmerman to allow a more upright orientation of the aircraft for VTOL operations.** *US Patent Office*

Above: **A complex post-war undercarriage and tail design for the Skimmer intended for VTOL operations. Convention runway take-offs and landings would remain optional.** *US Patent Office*

Above: **Internal features of the high-lift undercarriage proposed for an advanced version of the Skimmer.** *US Patent Office*

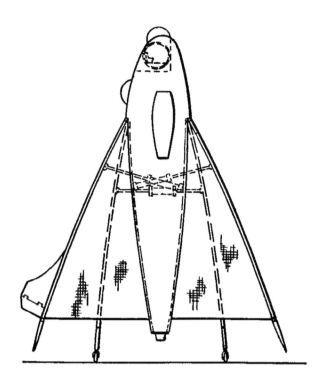
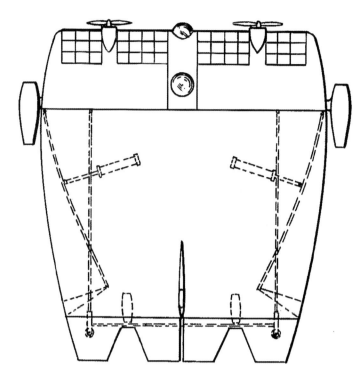

A design study produced by John Sullivan in 1954 to produce a more advanced VTOL version of the Skimmer using a complex landing gear system, a rotating cockpit and additional jet power. *Bill Rose*

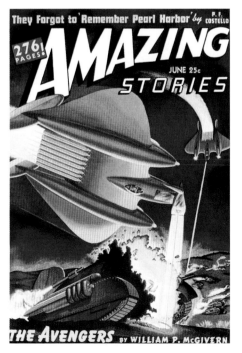

JET SKIMMERS?

Some months before the US Navy decided to proceed with development of the XF5U-1, the SF magazine Amazing Stories (June 1942, Volume 16, No 6) carried artwork on its front cover showing a very advanced rocket-powered Flying Pancake-type aircraft. Two of these orange aircraft are seen swooping down on futuristic tanks and attacking them with directed energy weapons.

It is inconceivable that the well-respected, Chicago-based artist Malcolm Smith (1910-1966) who produced this artwork had access to any classified documentation. However, it does appear that he tried to predict an advanced aircraft evolved from an Arup-type design. While rocket propulsion was never a realistic proposition, there have been claims that a purely turbojet-powered aircraft based on the Skimmer was built and tested in the immediate post-war period.

No documentation supporting such an aircraft exists, but it is possible that a prototype was developed in the black domain. Such an aircraft might explain some of the more reliable UFO sightings from this period, although I have no supporting evidence to present.

BOEING'S RIVAL PANCAKE

In early 1943, just a few months after Vought had been given the go-ahead to build two XF5U-1 prototypes, Boeing produced a detailed study for a rival STOL shipboard fighter with broadly similar capabilities, utilising an elliptical wing and a single tail fin with a rudder. Although not normally associated with fighters, Boeing's Seattle design office started out by proposing an initial one-man, low-speed, low-cost, proof-of-concept demonstrator called the Boeing B.396.

It was expected to perform like the Vought V-173 and would be used to obtain valuable data on the low aspect ratio design. Power for the B.396 would come from a centrally located 290hp (216kW) Lycoming 0-290 piston engine driving a fixed pitch propeller, and the engine would be mounted in a bay located partly below the fuselage. This would be enclosed by a shroud acting as a duct, to facilitate easy cooling.

The B.396 would be slightly smaller than the proposed fighter with a wingspan of 25ft (7.62m) and a wing area of 319sq ft (29.6m²) and the NACA 64,2-015 aerofoil proposed for the fighter. Gross weight was expected to be 1,400lb (635kg) and the aircraft would be constructed from wood/plywood and

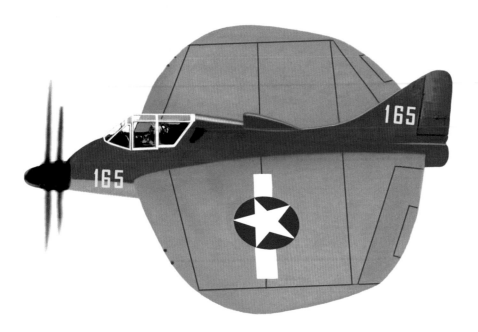

Left: **The Boeing B.390 was a proposed competitor to Vought's carrier-based STOL Skimmer. While the prototype Vought XF5U-1 was built and underwent initial testing, the Boeing design never left the drawing board. Although the B.390 would have probably lacked the performance of the Vought aircraft, its less complex design would have made it more reliable.** *Bill Rose*

WEYGERS DISCOPTER

In 1943, the American-based engineer and inventor Alexander George Weygers (1901-1989) completed several proposals for unusual flying machines called 'Discopters'. Although never built, these designs were close to the contemporary idea of a flying saucer and would be propelled by a ducted fan system.

His initial design was for a one-man flying disc with an overall diameter of 20ft 3in (6.1m) and a depth of 8ft 4in (2.5m). The craft would be piloted from a centrally positioned cockpit covered by a dome-shaped canopy and propulsion would take the form of two enclosed coaxial rotors turning around the cockpit area in contra-rotating directions. These would be driven by an unspecified engine directly beneath the cockpit. Air would be drawn in through a circular grille on the upper surface and directed downwards during flight, with flaps and vanes used for directional control. Weygers did not design this aircraft with an undercarriage, as it would rest on its lower surfaces while standing on the ground. It seems probable that Weygers expected the performance and range to be much better than a helicopter.

The one-man aircraft was followed by a series of designs for larger aircraft using the same general principles. These saucer-shaped concepts would have diameters of 50-100ft (15-30m), with the largest having three decks. Weygers considered various rotor arrangements, but favoured a contra-rotating design. There is little detail regarding the engine(s) driving the rotor system, which Weygers normally placed at the lower centre of the craft. One important issue that he did consider was the possibility of engine failure during flight and this would be a concern for all later disc-shaped designs that were totally reliant on their engines for flight. To deal with the concern, Weygers proposed a system of rocket motors that could be used to make an emergency landing. The Discopter was designed to rest on its underside and Weygers considered the aircraft suitable for use on water. That said, he acknowledged that a retractable undercarriage with wheels was desirable for ground handling.

Towards the end of 1943, Weygers applied

aluminium wherever possible. With a 12.5 UK gallon (56.7 litre) fuel supply, the B.396 was expected to have a two-hour endurance and an approximate maximum range of 250 miles (400km). The undercarriage was intended to be largely retractable, although clearly rather basic as it lacked covering doors. Maximum speed is not specified, but it would have been in the same range as the Vought V-173 and was not an important issue. The designation of this proposal suggests that the B.396 was conceived once some initial work was completed to establish the viability of a full-sized fighter aircraft.

After a period of testing, the first prototype semi-disc-shaped, single-seat Boeing B.390 fighter would replace the B.396. It would be powered by a single centrally located 3,500hp (2,610kW) Pratt & Whitney R-4360-3 Wasp 28-cylinder air-cooled radial engine. This would be connected to a forward reduction gearbox and a single counter-rotating assembly that used three-bladed propeller assemblies. Without this arrangement, it would have been very difficult to control the torque produced by the engine. Nevertheless, this design was mechanically simpler than the complex power transmission and propeller blade arrangement used by the Vought XF5U-1. Reliability should have been somewhat better, although the use of a radial engine in this manner may have caused some concerns about cooling.

Although the Boeing B.390 was expected to possess good STOL performance, its capability would never have matched the fully developed Vought F5U, which was designed to hover in a vertical attitude. Boeing suggested that the B.390 prototype would attain a maximum speed of about 425mph (683kph) at altitude and its rate of climb was estimated at 3,480ft per min (1,060m/min). The B.391 production version would be fitted with a two-stage supercharger, providing a top speed of about 452mph (757kph). One rather unusual feature was the ducting of the engine exhaust to the rear of the fuselage, which was expected to provide a small amount of added jet thrust. Range was calculated at about 1,000 miles (1,600km) and this could be extended with drop tanks.

Gross weight was set at 14,000lb (6,350kg) and the wingspan was to be 33ft 4in (10.15m). It was planned to use aluminium alloy for much of the B.390/391 airframe, with steel in critical areas. A fully retracting undercarriage would be fitted, plus arrestor gear, and visibility from the heavily armoured cockpit was expected to be reasonably good. Designated as the Boeing B.391, the production fighter would have been armed with four 20mm cannons and capable of carrying two 500lb (226kg) bombs or additional fuel tanks. Needless to say, this Boeing shipboard fighter never progressed beyond a paperwork study. However, it seems certain that this unusual aircraft could have been built, tested and put into production, but it was filed away and forgotten.

Top: **Original artwork showing one of the larger Discopter designs produced by Alexander Weygers in the early 1940s.** *Bill Rose*

Middle: **This illustration produced by Alexander Weygers in the late 1940s shows a future San Francisco where the Discopter has become the dominant means of air travel.** *Bill Rose*

Bottom: **Cross-section of a Discopter, showing the rotors and air ducting.** *US Patent Office*

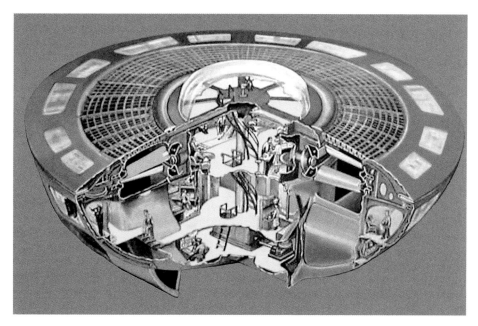

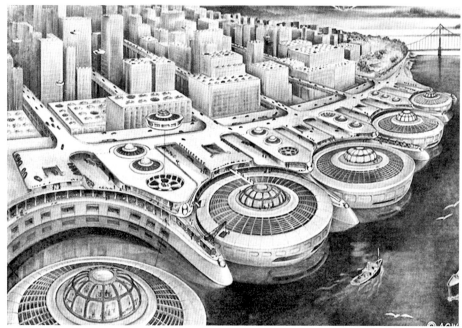

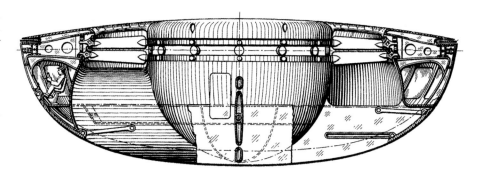

for a US patent, which was filed on 1 January 1944 (2377835). He also approached the USAAF and US Navy with his Discopter design, but the concept was considered too unusual for consideration at that time.

On 5 June 1945, the Discopter patent was published and although Weygers was working for US Army Intelligence, he still devoted much of his spare time to developing the design. In 1947 he produced drawings of a future San Francisco where a special port had been built for Discopter operations. Nevertheless, nothing came of these designs or plans, although they were undoubtedly influential and perhaps more realistic than Henri Coanda's earlier flying disc proposals.

Weygers was a multi-talented engineer, designer, artist and author with an exceptional command of different languages. He is often described as having been a modern Renaissance man and was clearly a unique individual. But the specifications for his patent and the available drawings suggest that it would have been difficult for anyone to make this intriguing concept work using prevailing technology. Major obstacles included adequate and reliable engine power, noise levels inside the aircraft and flight control. It has been said that the Discopter may have been influenced by Webb's 1918 Flying Machine and the 1933 concept produced by Paul Trenn in Germany.

DISCPLANES IN NAZI GERMANY

For some people, discplane experimentation in Nazi Germany has almost become a taboo subject. This is largely due to the mountain of complete nonsense that has been posted on the internet, often attempting to glorify the Third Reich.

In many cases, the claimed Nazi wonder-weapons are nothing more than politically motivated SF. There were no fleets of Nazi flying saucers that ended up at a secret base in Antarctica, no exotic anti-gravity propelled craft and absolutely no Nazi bases on the Moon. Nevertheless, most experts would agree that Nazi scientists were at the forefront

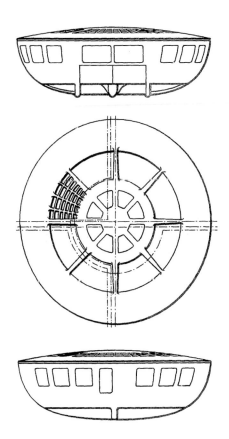

General appearance of the larger Weygers Discopter. *US Patent Office*

of aerospace development during World War 2. Germany was the first nation to field jet fighters in combat and launch long-range ballistic missiles against an enemy. German aerospace research would set the standard until the mid-1950s.

During the war, many unusual concepts with military potential were funded for further development and the following section is an attempt to present an honest outline of the handful that qualify for inclusion in this book. Large gaps exist in the detail and there are undoubtedly errors, but I must stress that these small-scale flying disc projects were experiments that generally led nowhere and would not have altered the outcome of World War 2 in any way.

HEINKEL-BMW PROJECT

Details of this controversial programme remain very hard to prove and are open to dispute. Furthermore, the distortion and spin generated by numerous over-enthusiastic writers and promoters of Nazi technology have damaged serious research, making it very hard to uncover fresh information. There are certainly enough pieces of the jigsaw to indicate that some parts of this story are true,

but authentic documentation remains elusive and may simply no longer exist.

Both Heinkel and BMW are said to have been actively involved in a secret aircraft project undertaken in Czechoslovakia. The origins of the story can be traced back to spring 1940, when Luftwaffe Hauptmann Rudolph Schriever (1909-1953) was assigned by the Reichsluftfahrtministerium (RLM: WW2 German Air Ministry) to work for the Heinkel Aircraft Company at Marienehe, Rostock.

Schriever gained an engineering degree in Czechoslovakia (possibly from The Czech Technical University in Prague) and at some point learned to fly, although relatively little is known about his early life. He began work in Heinkel's design office and showed a particular interest in the development of an aircraft that combined the best features of a helicopter and a fixed wing type. Apparently, his unusual ideas soon came to the attention of company chairman Ernst Heinkel (1888-1958) who encouraged Schriever to produce plans for a prototype. In early 1941, it was decided to proceed with the construction of a small prototype designed by Schriever.

Known as the Versuchs 1 (Versuchs – Experimental, or Version One), it also received the more popular name 'Flying Top'. Assembled in a small workshop, it is said that V1 had a diameter of about 2ft (61.0cm) and was propelled by an internal rotor driven by an electric motor or possibly a small two-stroke engine. Rather like the Zimmerman V-162, this was tested as a tethered demonstrator and by summer 1942, the results were sufficiently promising to gain approval to construct a much larger manned prototype called V2, with a VTOL capability.

Little is known about the V2 and it appears that almost every description of this vehicle is

Left: **This photograph of Rudolph Schriever is believed to have been taken in pre-war Prague and is almost certainly genuine.** *Bill Rose*

Above: **The object seen in this photograph is said to be the original Schriever 'Flying Top' demonstrator, although there is nothing to say it is anything more than a piece of unidentified industrial equipment.** *Bill Rose/Unknown*

incorrect or the product of invention. However, it does seem to have utilised a circular wing with a diameter of about 25ft (7.6m) that rotated around a central cockpit with a rear stabilising fin. Although there have been claims that gas turbine propulsion was used, this is simply guesswork and despite many years of investigation, the exact details remain obscure. Specialist contractors would have fabricated some of the major items for V2, with other parts being adapted from off-the-shelf components.

It is believed that assembly of V2 was undertaken at Heinkel's Marienehe facility in early 1943 and the craft was shipped to the company's site in Czechoslovakia for completion. This cannot be verified, so the exact course of events remains unknown. Furthermore, there are no records of an initial test flight, but assuming this prototype existed, it seems probable that Schriever attempted to fly it on at least one occasion.

Heinkel operated several facilities in Czechoslovakia during World War 2, with a discrete test facility at the huge Praha-Kbely complex (north-east of Prague). This was a major production and repair centre for various Luftwaffe aircraft and it remains one of the biggest aviation engineering facilities in Eastern Europe. The reason for moving this project to Czechoslovakia remains unknown, although it may have been due to the availability of skilled workers.

According to a report that was published in the February 1987 issue of the well-respected German aeronautical magazine *Flugzeug*, an unusual test flight was undertaken at the Praha-Kbely Airfield during August or September 1943. Apparently, the event was accidentally witnessed by 25 members of the locally based

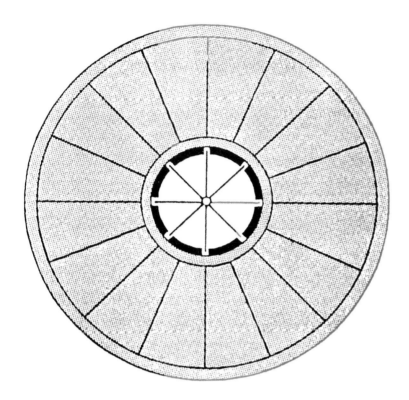

Possible appearance of the Heinkel-BMW V3 flying disc. *Bill Rose*

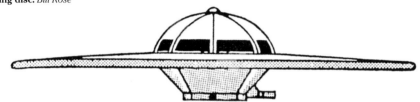

C-14 Flight School. The author of this article (who was a member of C-14) claims that the group observed a small disc-shaped aircraft being moved out of a hangar and started up. It then briefly hovered before undertaking a short flight just above ground level.

The aircraft (referred to as the 'Thing') was said to have had a diameter of about 20ft (6.1m). It was an aluminium colour with an outer rim that rotated, and was supported on the ground by four struts.

When it was realised that the C-14 group were watching this test, they were immediately ordered to leave the area, although they were aware of a second test flight taking place. Despite the apparent link, there is no absolute proof that this machine was the V2 prototype. Let me add that further reports of similar trials at Praha-Kbely Airfield cannot be substantiated.

By spring 1944, Schriever's team had fully relocated from Rostock to Prague and were undertaking further experiments at Heinkel's test facility within the Praha-Kbely complex. Work began on a new prototype called V3, which is thought to have been completed during autumn 1944. Schriever briefly discussed a design with a 47ft 4in (14.5m) diameter, which was powered by three gas turbines. This may have been V3, although there is no verifiable documentation to check.

Such an aircraft combining the capabilities of a fixed wing aircraft and a helicopter would have been regarded as extremely useful by the RLM, who required fighter aircraft that could operate from improvised sites, with the ability to engage high-flying bombers such as the anticipated USAAF B-29. With airfields under continual attack from Allied bombers, there were already frantic efforts to develop rocket-powered point defence interceptors that could operate from dispersed sites. This led to the Bachem Ba 349A Natter (Adder), which was rushed into service as the war ended but never saw combat.

Unfortunately, nothing is known about the V3, which may have simply been an engineering mock-up. If it was built as a functioning prototype there are no documents to show that it was test-flown, despite some claims to the contrary and bogus evidence posted on the internet. Regardless of this, there

Far left: **Dr Giuseppe Belluzzo, the Italian turbine specialist and right-wing politician who claimed involvement with a wartime flying saucer project that was perfected by the Germans.** *Bill Rose*

Left: **The elusive Richard Miethe, possibly photographed at a pre-war rocket test site.** *Bill Rose*

were changes made to the way the project was being run and in late 1944 the SS are thought to have taken direct control of the Prague Project under the direction of General Dr Hans Kammler.

According to information from several different sources, there was an expansion of the project, which involved BMW. Several highly qualified specialists joined the team and they included the enigmatic Dr Richard Miethe (1912-?), who is said to have worked on the development of V-1 and V-2 missiles. He is believed to have been a member of staff at the Specialist Advanced Propulsion Section of BMW Bramo at Berlin-Spandau, which was directly associated with the Peenemünde rocket site.

Another claimed member of this expanding group was Klaus Habermohl, a BMW Bramo engineer. BMW's Specialist Advanced Propulsion Section was headed by Dr Helmut Zborowski (1905-1969), who may have taken overall control of Schriever's flying disc project from Berlin. While this project originated at Heinkel, the involvement of BMW would seem to confirm the intended use of a new propulsion system that was in development.

Also said to have been connected to Schriever's project (at some stage), was Dr Giuseppe Belluzzo (1876-1952), an Italian turbine specialist who is thought to have worked as an external consultant. He would later claim to be the originator of a flying disc design that was perfected by the Germans. There is no evidence to indicate that the aerodynamicist Henri Coanda was recruited for this project, and claims to this effect appear completely false.

Heinkel had pioneered the gas turbine, but

the company was regarded as an airframe manufacturer by the RLM and discouraged from starting jet engine production. Because of this, BMW and Junkers took the lead, although Heinkel's Hirth Motoren engine company based at Stuttgart-Zuffenhausen would have commenced production of advanced gas turbines if hostilities had lasted longer.

It is said that this unusual alliance between Heinkel and BMW came about because an aircraft based on V3 would provide the ideal platform to test a new experimental type of jet engine. This may have been an early form of Radial Flow Gas Turbine (RFGT), which was shaped like a disc with an outer rotor, making it well suited to a circular-shaped aircraft. Unusual gas turbine concepts had already been considered by the Detroit-based engineer John Heinze during the late 1920s and René Leduc in the late 1930s, so it is possible that a BMW design may have evolved from this work.

V3 had probably been abandoned by this time, but several paperwork concepts followed (V4-V6), leading to the construction of a prototype called V7. This design is attributed to Miethe and seems to have been developed from the outset to take full advantage of the BMW prototype RFGT engine. The aircraft had an overall diameter of about 60-70ft (18.3-21.3m), with accommodation for a crew of two and perhaps enough space for a third person.

It has been suggested that Reichsmarschall Hermann Goering visited the site on several occasions to inspect V7, but this remains uncorroborated. Another visitor to the Praha-Kbely project was Luigi Romersa (a twenty-seven year old Milan-based journalist who worked directly for

Mussolini as an advisor). During a 2004 TV documentary, he claimed to have been shown a prototype flying saucer on the direct authority of Hitler.

V7 is thought to have been completed in early 1945 and Georg Klein, who worked as an engineer on the project, claimed to have seen V7 flown for the first time in February 1945. Major (Reserve) Rudolph Lusar, author of the authoritative mid-1950s book *German Secret Weapons of the Second World War*, also mentioned V7, suggesting that Schriever and Habermohl flew this aircraft. Lusar is understood to have worked for the German Patents Office as a consultant engineer, so he was well placed to discuss advanced wartime projects.

There are several reports that V7 achieved supersonic speed during its first (and possibly only) test flight, reaching an altitude of 40,000ft (12,192m). This is too ridiculous to take seriously. If V7 was flown, it will have followed at least 100 hours of engine testing and is unlikely to have been anything more than one or two brief tethered lift-offs.

What happened to the prototype at the end of World War 2 is equally ambiguous. It is possible that the Waffen SS destroyed V7 when Field Marshal Wilhelm Keitel ratified the terms of surrender on 9 May 1945. It is also possible that Soviet forces secured V7, research documentation, equipment and some of the personnel who worked on this project.

There are also claims that the Prague team were developing a supersonic flying disc known as V8. Apparently, this was another Miethe design that utilised an advanced boundary layer control system which provided a Mach 3 capability. If this SF design existed in any form, it was probably nothing more than a drawing.

Following Germany's surrender, Rudolph Schriever took his family to Bremerhaven-Lehe, where they lived with his parents. He was lucky enough to secure a job driving for the US Army and his duties mainly involved the delivery of publications such as the Stars & Stripes to local military camps.

During the late 1940s, UFO sightings started to grab headlines around the world and Schriever attempted to have his account of the

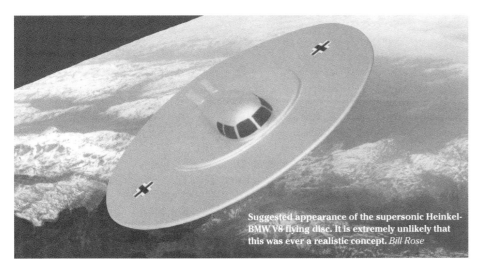
Suggested appearance of the supersonic Heinkel-BMW V8 flying disc. It is extremely unlikely that this was ever a realistic concept. *Bill Rose*

flying disc project taken seriously. This finally happened on 30 March 1950, when *Der Spiegel* published an article about him, which included a rather fanciful drawing of a flying saucer that was called *Der Schrieversche*

Below: **A drawing made by Giuseppe Belluzzo in 1942, which was reproduced in Italian newspapers during 1950. It appears to show the turbine propulsion system for a flying disc design. Belluzzo claimed to have been involved with work being undertaken by German engineers to develop a flying disc aircraft.** *Bill Rose*

Below right: **This drawing shows the main features of a flying disc-shaped vehicle, which Dr Giuseppe Belluzzo claimed to have produced in the early 1940s.** *Bill Rose*

Flugkreisel (The Schriever Flying-gyroscope). The craft in this illustration appears to be a product of the artist rather than Schriever and it is remarkably similar to a concept seen in the US SF magazine *Amazing Stories* for July 1943.

Newspaper articles followed this feature, Schriever insisting that the saucer documentation and blueprints had been taken by the Soviets, who were now building advanced reconnaissance craft that accounted for many recent UFO sightings. Despite their unpublicised concerns, the American and British authorities showed no interest in Schriever's revelations and many researchers suggested his story was fantasy and that he was unbalanced. Throughout 1951

and 1952, Schriever continued to generate interest in his flying disc story. Then he was killed in a rather a mysterious traffic accident on 16 January 1953. Some present-day conspiracy theorists have suggested there was more to this event than meets the eye.

Exactly what happened to Miethe at the end of the war remains unclear, although the author was advised from a very reliable source that he escaped from Czechoslovakia in early May 1945 and eventually contacted American intelligence. He was then held along with other important scientists, including Dr Wernher von Braun and a large group of his rocket scientists. T-2 Technical Intelligence Personnel interrogated them at length but, beyond this, the details start to become unreliable. Evidence strongly suggests that Miethe was sent to America, arriving in Fort Bliss. He was then moved to Wright Field AAF, which was the main location for evaluating captured German aeronautical technology.

If we are to believe an article that appeared in the French newspaper *France-Soir* on 7 June 1952, Miethe returned to Germany where he became involved with a group of German rocket scientists headed by Kurt Fuellner and Herman Plitzken. They were described as V-weapon specialists who travelled to Egypt where attempts were made to develop rockets for the military.

Apparently there was a very unpleasant dispute with local officials and most of the

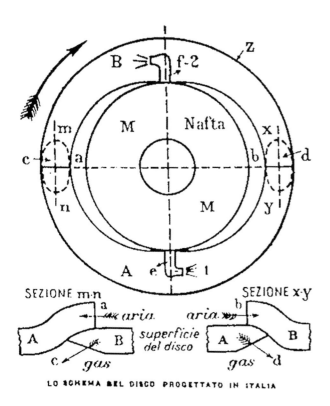

LO SCHEMA DEL DISCO PROGETTATO IN ITALIA

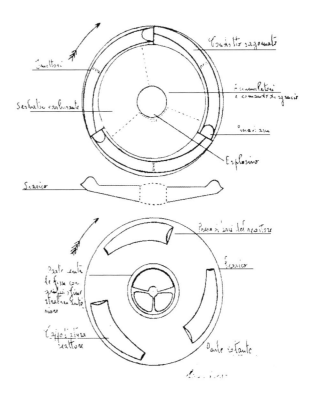

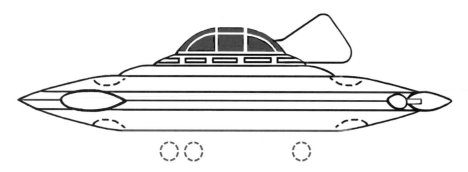

Left: **Side view of the flying saucer design attributed to the Italian engineer Francesco de Beaumont. This design may have been derived from work undertaken by German engineers at Prague.** *Bill Rose*

group hastily returned to Hamburg. One of the scientists left behind was forty-year-old Dr Richard Miethe, who reportedly fled to Tel Aviv, Israel. Some parts of this story sound accurate while other details seem intentionally deceitful. Nevertheless, in this *France-Soir* report, Miethe revealed that he had worked on a flying disc aircraft called V7 during the war.

What happened to Miethe after this is unknown and he simply vanished off the radar. Many ufologists have been sold on the idea that Miethe secretly participated in the 1950s Avro-Canada flying disc programme and the origin of this suggestion can be traced to Major Rudolph Lusar's book *German Secret Weapons of the Second World War*. However, this appears to be completely speculative and the author's research in Toronto failed to find any kind of evidence that he ever visited the country, or directly participated in the Avro-Canada project.

The post-war fate of propulsion specialist Klaus Habermohl is equally vague. Several reports claim he was taken prisoner by the Russians and put to work at a secret and still unknown Opytnoe Konstruktorskoe Byuro (OKB or Experimental Design Bureau). Whether or not this is true remains unknown, because there is no evidence whatsoever that Habermohl actually existed.

Dr Belluzzo is unquestionably the best-known scientist associated with the flying disc development during the 1940s. He was a highly respected engineer and designer who produced a flying disc proposal during the early 1940s. As for his alleged involvement in or connection to the Heinkel-BMW Project at Prague, this remains somewhat questionable. Historical records show that Belluzzo was in Milan during the city's liberation and was detained by the Comitato di Liberazione Nazionale (CLN: National Liberation Committee). After his release, Belluzzo became associated with the Constituent Assembly and his earlier support for the Fascist regime appears to have been tolerated.

It is interesting to note that Belluzzo's revelations in 1950 about his participation in a wartime German flying disc project appeared in the newspaper *Giornale D'Italia* just a matter of days before Schriever's *Der Spiegel* feature. There is no obvious connection or reason for this coincidence. Exactly how far his claimed links to advanced German aviation projects extended remains unknown and Belluzzo died in Rome on 21 May 1952.

Two very interesting flying disc designs were produced in Rome during the early 1950s and it has been suggested to the author that Belluzzo had an initial involvement with the production of these concepts. They are attributed to the inventor and designer Francesco de Beaumont and his first aircraft may have incorporated certain features associated with Schriever's early work.

This design utilised a rotor system to provide VTOL and horizontal flight that was driven by four separate gas turbines. The role envisaged for this flying disc remains unclear, although de Beaumont mentions a high-speed capability and the possibility of piloting the aircraft by remote control.

A second flying disc design produced by de Beaumont may have been loosely based on research undertaken by Belluzzo for the Heinkel-BMW programme at Prague. Whether this design relates to the V7 remains unknown. For propulsion, it utilised a large rotor assembly turning around the circumference of the vehicle, which would be driven by four equally spaced centrifugal flow jet engines in pods. As an alternative to the jet engines, pulsejets or ramjets might be used.

The rotor assembly would provide

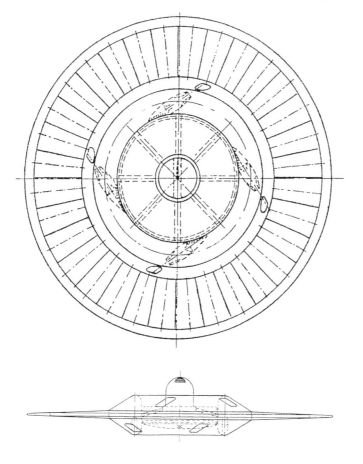

Left: **This de Beaumont rotor-wheel flying disc design may be based on earlier work undertaken by Dr Giuseppe Belluzzo and German engineers in Prague.** *French Patent Office*

Flying Saucer Technology

gyroscopic stability during flight. In addition to a full VTOL and hover capability, a high-subsonic speed was anticipated in level flight. Constructed of steel and aluminium alloy, the aircraft is built around a central cockpit section fitted with a dorsal stabilising fin, sometimes seen in drawings of Schriever's flying discs. It was planned to support the aircraft on a fully retractable undercarriage, and de Beaumont suggested this craft might also be flown by remote control. Many aspects of this concept appear to be rather unsatisfactory and are not clearly explained in available documentation. Needless to say, nothing came of the design.

Of the remaining people who are said to have been involved with this project, Dr Helmut Zborowski established a successful design consultancy in France during 1947 called Bureau Technique Zborowski (BTZ). They worked for a number of defence contractors such as SNECMA, designing advanced VTOL fighter concepts that were derived from wartime research.

The fate of SS General Dr Hans Friedrich Karl Franz Kammler is unknown. It has been said that he was shot dead during May 1945, but it is equally possible that he was helped to escape from Germany. An outsider claiming links to the Prague project was Joseph Andreas Epp (1914-1997). He grew up in Hamburg and after leaving school worked as an engineer in several different capacities. In 1936, Epp joined the Luftwaffe, becoming an aircraft maintenance technician.

By 1939, he was working on his own designs for a new type of aircraft that would combine the best features of a fixed wing aircraft with those of a helicopter. Epp described his concept as the Helioplane and designs were forwarded to Armaments Minister Ernst Udet. Epp heard nothing more, but claims to have discovered that development of his discplane was under way at Prague. The project had started in 1943 and a prototype was completed but ran into problems. A second version followed and this was test-flown in February 1945. Epp had no involvement with this work and insisted that Schriever's design was amalgamated with his and developed further by three SS appointed specialists.

In 1943, Epp began work on the design of a new aircraft that he called the Omega Diskus. This utilised eight ducted lift fans and two Pabst ramjet engines, which were under development for the proposed Focke-Wulf Triebflügel (Thrust Wing) VTOL fighter. They were supported by an overhead two-blade assembly and rotated around the circumference of the disc. The Omega Diskus

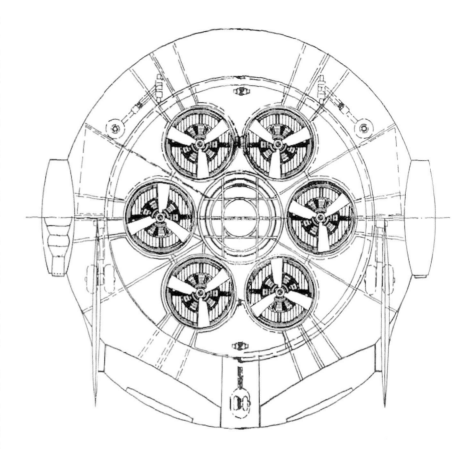

was 62ft 4in (19m) in diameter and 72ft (22m) when allowing for the ramjets. There was a large centrally positioned 13ft (4m) circular cockpit and a fully retractable four-wheel undercarriage.

After the war, Epp attempted to interest the British and Americans in his unusual design, but there were no takers. He then claimed to have been recruited by the KGB to work on a prototype flying saucer and was given responsibility for designing this aircraft's control system.

Epp insisted that this saucer-shaped aircraft was built and tested in East Germany, but he became disillusioned with the programme and escaped back to West Germany. Following further unsuccessful approaches to Western aircraft contractors, Epp relocated to Spain where he lived until 1951, when he returned to West Germany. Despite being unable to secure any financial backing for his aviation projects, Epp continued to work on various flying disc designs, and by 1958 he had constructed another demonstrator that was test-flown.

Nothing came of this and Epp would spend a large part of the next four decades trying to have his designs and ideas taken seriously. Epp always insisted that UFO sightings around the world were secretly developed machines based on work initially

Above: **Based on an original drawing by Andreas Epp, this flying disc is said to have been developed in Soviet-occupied East Germany during the immediate post-war years. Throughout his later years, Epp claimed to have been the originator of prototypes built under the direction of Schriever at Prague and the later VTOL reconnaissance craft shown here.** *Bill Rose*

Below: **Luftwaffe pilot Andreas Epp, who claimed to have submitted designs later used to develop the flying discs built at Prague.** *Bill Rose*

undertaken in wartime Germany. Epp died alone in September 1997, having lived for some time on State benefit. There are many engineering issues with his aircraft designs and it is hard to accept Epp's rather casual involvement with a classified Soviet aircraft project in East Germany. While this is an interesting and often contradictory story, it is impossible to verify the details.

Finally, this section would not be complete without mentioning Heinrich Fleissner (1904-?), a hydraulics engineer who worked on wartime rocket programmes at Peenemünde. He claimed to have been involved with the development of a secret flying disc that was shipped to Peenemünde for assembly and testing. This unusual craft was called the Düsenscheibe (Nozzle Disc) which reflected its propulsion system; Fleissner said that this prototype was test-flown a couple of weeks before the war ended.

Apparently, the Düsenscheibe was a flying disc of substantial size taking the form of a central cockpit surrounded by a rotating disc powered by several turbojet engines. However, Fleissner made many unrealistic claims for the performance of this VTOL aircraft, which did nothing for his credibility.

The rotating disc surrounding the central cabin was to all intents and purposes a type of ramjet turbine that would be started using small rockets. It provided gyroscopic stability, with directional thrust generated behind and below the disc. The central cabin was electromagnetically stabilised to maintain its

position, and directional control of the disc was achieved with the use of ducts containing vanes and flaps. The double convex disc would be supported on the ground by three retractable struts with inflatable balls attached that would allow landings on water.

Fleissner continued to evolve this design, applying for a German patent on 27 March 1954. His new concept was called a 'Rotating jet aircraft with lifting disc wing and centrifugal tanks' By the following year, he had secured a deal with Paul Tobeler, a Californian businessman, who took a half share of the design in exchange for filing a US patent and attempting to sell it to a US aircraft contractor. The patent application was presented on 28 March 1955 and for reasons that remain unknown it was not published until June 1960 (2939643). It is possible that Fleissner's design was considered too similar to the classified USAF Silver Bug project and this raised a number of security issues. But Tobeler was unable to sell Fleissner's design and the idea was finally abandoned. In 1980, Fleissner told his story to the German newspaper *Augsburger Neue Presse*, which described him as the 'father of the flying disc'.

Again, there does appear to be substance to some parts of this story. There is the connection to missile development at Peenemünde, although some of details do not stand up to close scrutiny and may have been said simply to impress anyone willing to listen. Was the alleged Düsenscheibe prototype actually V7 and could it have been shipped from Prague to Peenemünde for final assembly and flight-testing? Was Fleissner's post-war design built around an unusual type of experimental engine

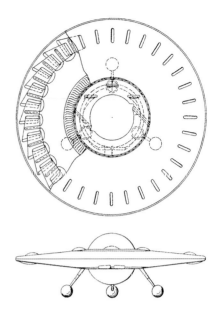

Above: **This drawing shows the main detail of Heinrich Fleissner's flying disc design, which may have been based on wartime research. During the 1950s he tried to sell his plans to an American aircraft manufacturer, but nothing came of the concept.** *US Patent Office*

produced by BMW? Unfortunately, the true story may never be known as there is a complete lack of hard evidence to support any claims that flying discs were developed by Heinkel, BMW or various scientists connected to Peenemünde.

It is easy to dismiss everything in this section as nonsense that belongs in the realms of SF, although the lack of historical evidence is hardly unique. In the closing days of World War 2, huge quantities of documentation were lost, intentionally destroyed or retrieved by Allied Intelligence and immediately classified. Some of this information remains very sensitive and will probably never reach the public domain, such as formulations for unusual chemical weapons or advanced cipher systems. I should also briefly mention that a serious fire at the Avro Aircraft Company in England during the late 1950s destroyed almost every company record, including all details of advanced military projects. Therefore, a lack of documentation does not automatically indicate that every account of a particularly obscure aviation project is bogus.

Whatever the truth might be, it is interesting to note that by the start of 1953 every member of the Prague Project apart from Zborowski had either died or disappeared. It is tempting to think there might be a bigger story that remains untold. It would certainly make a good movie.

Below: **A photograph of Epp's Omega Diskus model aircraft, showing the eight ducted fans and upper rotor arrangement.** *Bill Rose*

SACK DISCPLANES

Perhaps the best-known circular-winged German World War 2 design was not only low-tech but did not even work properly. The existence of this propeller-driven aircraft, known as the AS-6, was only proven when a series of black and white photographs surfaced during the early 1970s.

The aircraft's origins can be traced to Germany's first National Contest of Aero Models with Combustion Engines, held at Leipzig-Mockau on 27-28 June 1939. One of the competitors was Arthur Sack (1900-1964), a landowner and model aircraft constructor who entered an unusual monoplane he called the AS-1. This circular-winged aircraft had a reported span of about 50in (127cm), a length of 62in (157cm) and a height of 26in (66cm). AS-1 was propelled by a small Kratmo-30 engine driving a two-blade propeller with a diameter of 23.5in (59.7cm) and the aircraft's weight was 9.9lb (4.49kg). An undercarriage was fitted, consisting of two forward struts with wheels and a tail skid.

The reasons why Sack built a circular-winged model aircraft are unknown. Perhaps it was simply the unusual shape of the wing that appealed to him, or he may have taken note of the STOL designs built in America. When the time arrived to demonstrate his model, Sack was unable to make powered

flights due to difficulties with the Kratmo-30 engine. In desperation, he resorted to several hand launches which showed the aircraft's ability to glide quite well. This was sufficient to catch the attention of Air Minister Ernst Udet who was present at the contest and was sufficiently impressed with the unusual design to encourage further development.

With modest funding from the RLM, Sack spent the following three years steadily refining his design and is believed to have received a certain amount of guidance from the aviation designer Dr Alexander Lippisch. Sack

continued to build model aircraft and AS-1 was followed by AS-2 to AS-5. These are all thought to have been similar in size to the original model, but there was a steady improvement in aerodynamics, accompanied by a significant weight reduction.

AS-5 had flown successfully in early 1943 and RLM officials had been sufficiently impressed to approve the construction of a full-sized manned prototype. This would be known as the AS-6 Versuchs 1 (V-1).

Mitteldeutsche Metallwerke (which designed and built assault gliders in

Right: **General appearance of the Sack AS-5, which was the most sophisticated version of this series of small model aircraft.** *Bill Rose*

Below: **Air Minister Ernst Udet was directly responsible for encouraging further development of Sack's flying saucer aircraft.** *Bill Rose*

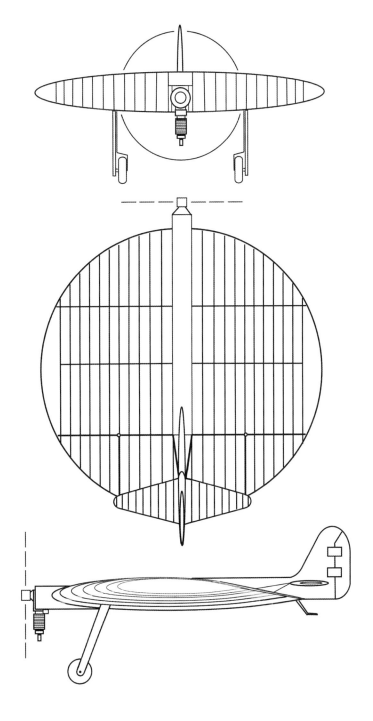

Left: **Arthur Sack's AS-6 V-1 circular-winged experimental aircraft at Brandis Airfield with its engine covers removed.** *Bill Rose*

association with Gothaer Waggonfabrik) were assigned the task of fabricating the basic AS-6 airframe at their Leipzig facility. The work commenced in autumn 1943 and the airframe was built almost entirely from wood (mostly plywood) and covered with fabric. By the start of 1944, Mitteldeutsche Metallwerke had completed the airframe and it had been transported to Brandis Airfield for completion by Luftwaffe engineers. The RLM had set a number of stipulations for this project, one of which was the use of salvaged components whenever possible. Most of the parts used would come from wrecked Messerschmitt aircraft at the airfield.

Power for the AS-5 was supplied by an Argus As 10C-3 240hp (179kW) inverted V-8 engine recovered from a Bf 108 Taifun utility aircraft and equipped with a two-blade wooden propeller. All the cockpit components

came from a crashed Bf 109B fighter, and another scrapped Bf 109B provided the undercarriage. This was attached in a non-retractable position and a small tail skid was fitted, which was later replaced by a wheel.

When the aircraft was finally completed, AS-6 had a wingspan of 16ft 5in (5m), a wing area of 211sq ft (19.6m²), an overall length of 21ft (6.4m) and a height of 8ft 5in. This was a very compact aircraft, with a gross weight of approximately 1,984lb (900kg). There are no available documents showing anticipated performance for the AS-6, but a conservative estimate would suggest a maximum speed in excess of 175mph (280kph) and a ceiling better than 15,000ft (4,572m). Engine and taxiing trials commenced towards the end of March 1944, undertaken at Brandis by Rolf Baltabol who was the chief pilot for ATG (a Junkers subcontractor).

It was apparent from the outset that AS-6 was poorly constructed and handled badly. Baltabol made five fast runs along the 4,100ft (1.25km) runway and found the aircraft very difficult to control. During the final run, the landing gear was damaged and this brought the trials to an end for several weeks. As a consequence, Baltabol insisted that a number of modifications were made to the rudder and ailerons. The repairs and modifications had been completed by 16 April 1944 and the aircraft was prepared for another attempt to get it into the air.

Conditions are said to have been good and although the handling was better, Baltabol was unable to take off. It is believed that during a second attempt, the aircraft began porpoising towards the end of the runway and the propeller was damaged when it made contact with the ground. Repairs followed and Sack tried to generate more lift during take-off by having the undercarriage legs moved rearward by about 8in (20.3cm) to raise the aircraft's nose.

Nevertheless, Baltabol was concerned about the increased likelihood of AS-6 tipping forward during take-off and the propeller hitting the ground. Sack had actually considered moving the undercarriage struts back by another 8in (20.3cm), but Baltabol made it very clear that he was unwilling to continue testing the aircraft if this happened. Sack reluctantly accepted his concerns and dropped the idea, although a range of other modifications were made. These included the installation of brakes from a scrapped Junkers Ju 88 bomber, new larger control surfaces and the carriage of 150lb (70kg) of ballast in the rear fuselage.

The next attempt to fly AS-6 was equally disastrous. Baltabol managed to briefly lift off from the runway but found the aircraft almost impossible to control, having to bank hard to the left. He managed to put the aircraft down but told Sack that he believed the wings were too short to compensate for the engine's torque. His involvement with the project was over. No further flights were attempted and there was reluctant acceptance that the aircraft needed to be wind tunnel tested and fitted with a more powerful engine. AS-6 was

Left: **Prior to the widespread reproduction of this image, the existence of the Sack AS-6 V-1 was dismissed as bogus. Some heavily retouched versions of this image appear to have been altered to enhance the aircraft's circular appearance.** *Bill Rose*

now moved into a hangar while major changes took place at Brandis.

The airfield had become the centre of operations for I/JG 400, equipped with the new Messerschmitt Me 163B rocket fighter. Their primary function was the daylight interception of USAAF and RAF bombers attacking the massive Leuna-Merseburg synthetic fuel plant operated by I. G. Farben. This was one of several industrial sites vital to the German war effort.

After their arrival at Brandis, some members of I/JG 400 discovered Sack's aircraft and Oberleutnant Franz Rössle (regarded by his fellow officers as something of a daredevil) insisted on trying to fly it. It is unclear if Sack gave his approval, but Rössle wasted no time having the aircraft fuelled and raced down the runway. But he was unable to control AS-6 and ended up bouncing across some rough ground. This damaged the undercarriage and AS-6 was towed back to the hangar.

In early 1945, a group of Allied warplanes succeeded in penetrating the airfield's defences and caused widespread damage. The hangar storing AS-6 was hit and the aircraft was almost completely destroyed. Its remains ended up on a scrap heap and all traces of AS-6 V-1 had vanished when the US Army's 9th Armoured Division arrived at Brandis on 20 April 1945. According to reliable sources, Sack met with engineers from Messerschmitt A.G. during early 1945 to discuss his ideas for a more advanced circular-winged aircraft. This was apparently called AS-7 V-1 or the Bussard (Buzzard).

Sack proposed the combination of a Messerschmitt Bf 109K-4 (the final production version of this fighter) with a newly devised circular wing. Equipped with a DB 605 ASCM/DCM V-12 engine rated at 2,000hp (1,491kW), driving a contra-rotating propeller unit, AS-7 might have benefited from good STOL performance and a high maximum speed. Although interest had largely switched to jet-powered interceptors, the Luftwaffe would have been attracted to a modified production aircraft that had been optimised for use at improvised sites such as short stretches of autobahn. Messerschmitt are said to have provisionally assigned the designation Me 600 to Sack's AS-7 V-1, but whether or not this is true, nothing came of the project.

FOCKE-WULF SCHNELL FLUGZEUG–ROCHEN

On 1 January 1924, Focke-Wulf Flugzeugbau A.G. was formally created. Although little known at that time, it would become one of Germany's most famous aviation companies within two decades.

Professor Heinrich Karl Focke (1890-1979) was one of the three founders of Focke-Wulf, but political pressure and his preoccupation with autogyros and helicopters would result in Focke leaving the company during 1937. He then started a new company with the pilot Gerd Achgelis, called Focke-Achgelis Flugzeugbau, which was based at Hoyenkamp near Delmenhorst. Their sole purpose would be the design and construction of advanced helicopters.

The company was run directly by Professor Focke and their first prototype was the Fa-61, which evolved into the Fa-223 Drache (Dragon) military helicopter. This entered limited service for air crash recovery and at least one highly classified operation was undertaken. Focke-Achgelis also developed and built the Fa-330 Bachstelze (Wagtail) autogyro for tethered use from a U-boat. These tiny one-man aircraft were utilised in small numbers by the Kriegsmarine, although their value remains questionable.

Many innovative engineering features were produced by Professor Focke during the existence of his company, such as the turboshaft, which is a universal feature of modern helicopters. Focke-Achgelis also developed plans for several intriguing concepts such as the tilt-wing Fa-269 and the very advanced Schnellflugzeug (Fast Aircraft)

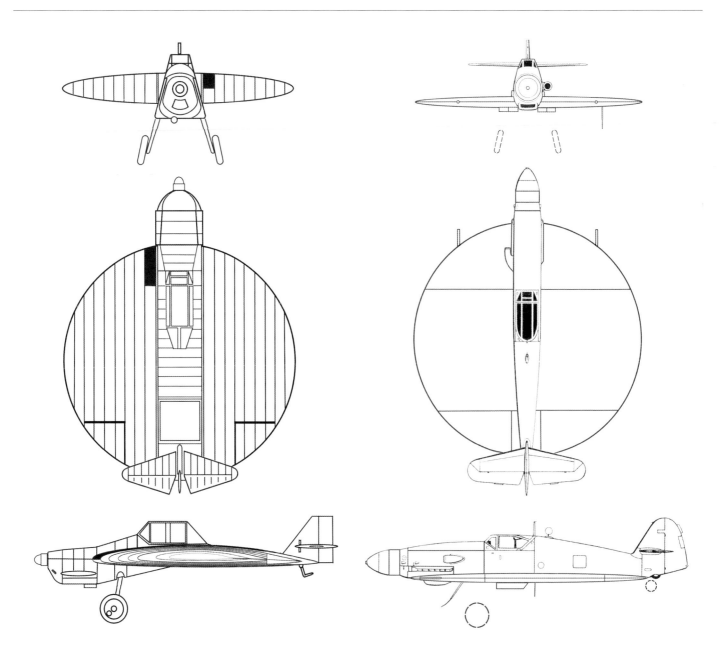

VTOL fighter. This design originated in the late 1930s, when Heinrich Focke proposed a circular-bodied aircraft that would take off and land vertically using a large centrally positioned duct containing two contra-rotating propellers designed to eliminate torque problems.

When gas turbine development had advanced sufficiently for application to this design, Focke envisaged a single jet engine behind the duct that would drive both propellers with a turboshaft. The engine's exhaust would be ducted through two separate trailing edge outlets. During VTOL operation, the airflow through the lower part of the duct would be directed by a series of louvred shutters to provide positive control in hover and low-speed flight.

The upper opening was envisaged as a conformal grille. After achieving horizontal flight, the lower louvred shutters would be angled rearwards, with additional thrust provided by the engine exhaust. This would be ducted to two outlets at the trailing edge and the system would include an elementary form of afterburning to generating additional thrust when required. It has been suggested that the output of each exhaust could be varied for additional control of yaw at low speeds. The entire underside of the fuselage would generate lift and Focke believed that in the event of an engine failure, a horizontal landing could be made with the louvres closed.

The one-man cockpit would take the form of a forward-protruding nacelle with the pilot sitting in a normal upright position. It is

Above left: **The Sack AS-6 V-1 aircraft.** *Bill Rose*

Above: **This drawing indicates the possible appearance of an advanced fighter intended to follow the Sack AS-6 V-1. Built by Messerschmitt, it has been suggested that this aircraft would be designated as the Sack AS-7 or the Me 600 and given the name Bussard (Buzzard). Sack is believed to have met with Messerschmitt engineers towards the end of World War 2 to discuss this idea, but it seems unlikely his proposals advanced much further.** *Bill Rose*

assumed that an ejector seat was envisaged, although no details can be found. A single tail fin and rudder would improve stability and provide additional flight control, while elevons and flaps were located along the trailing edge. There would be a fully retractable undercarriage, with two substantial forward struts and wheels and a tail wheel. On the ground the aircraft would be raised by about 18° at the front and in this position the rotors would be level with the ground.

There are no details of dimensions, but a very rough estimate suggests a length of about 59ft (18m) and a span of about 39ft (12m). This would indicate that the large central duct was expected to be about 20ft (6m) in diameter. No weights or performance figures are specified. Although this was a Focke-Achgelis design, it is thought that work on this project may have involved one of Kurt Tank's Special Project teams at Focke-Wulf Aircraft. Wind tunnel testing was undertaken and the design continued to evolve, although it was not given priority and the pace was slow. The exact role envisaged for this aircraft remains unclear, but it was probably seen as a fighter capable of operating from improvised sites. Armament would have taken the form of several cannons or possibly air-to-air rockets.

The Schnellflugzeug was not the only concept of this type to emerge during the 1940s. On 19 April 1945, American designers Charles Neumann and Hugo Baca filed a US patent for a 'Fluid controlled and Propelled' VTOL aircraft which utilised a central duct with contra-rotating propellers and louvred shutters for directional control.

Horizontal propulsion would be achieved by ducting the airflow to the rear of the aircraft, the design being apparently influenced by the Discopter proposed by Alexander Weygers. While their two-seat, spade-shaped aircraft shares similarities with Professor Focke's Schnellflugzeug, this appears to be nothing more than a coincidence. The Neumann-Baca patent was published in early 1949, but it progressed no further.

A somewhat similar design was produced in June 1948, by George Harold Naught (1902-1990), an inventor who lived in Rushville, Illinois. His concept was a circular saucer-shaped craft with a central duct containing a four-bladed propeller. Louvred shutters regulated the airflow and Naught proposed a rear-mounted jet engine for horizontal flight. Horizontal flight would be controlled by a tail fin with a rudder and two short wingtip fins with ailerons. Naught's aim was to produce a more advanced helicopter for use as a light utility aircraft and although the design was never built, it generated professional interest.

This was followed by a US patent application filed on 9 March 1953 by the Goodyear Aircraft Company of Ohio for a

design called a Convertible Aircraft, which utilised some features of the Discopter and aspects of Professor Focke's Schnellflugzeug. Designed by Robert Ross, the Goodyear Convertible Aircraft was built around a central duct containing contra-rotating variable pitch four-blade propellers tilted forward by 7° to produce lift for VTOL operations. Louvred shutters would be used to direct the airflow and these would close in horizontal flight with the airflow directed rearward. The powerplant would be positioned directly below the propeller assembly, with exhaust gas being removed by the airflow. Flight control would be achieved by a control surface with a three-fin tailplane and central rudder. There would also be triangular ailerons at the corners of the trailing edge. The Convertible Aircraft

was equipped with a fully retractable tricycle undercarriage and the forward cockpit nacelle might have been intended for a two-man crew.

The role Goodyear envisaged for this aircraft is unclear, but it seems likely they were considering the design for a naval carrier or warship-based operations. Performance was expected to be superior to any helicopter and perhaps this proposal was for observation or anti-submarine duties? Goodyear are not usually associated with the design or production of aircraft, but

Below: **Principal features of the wartime Focke-Achgelis VTOL design. This project was resurrected by the reformed Focke-Wulf Company in the early 1950s, but never progressed beyond wind tunnel testing.** *Bill Rose*

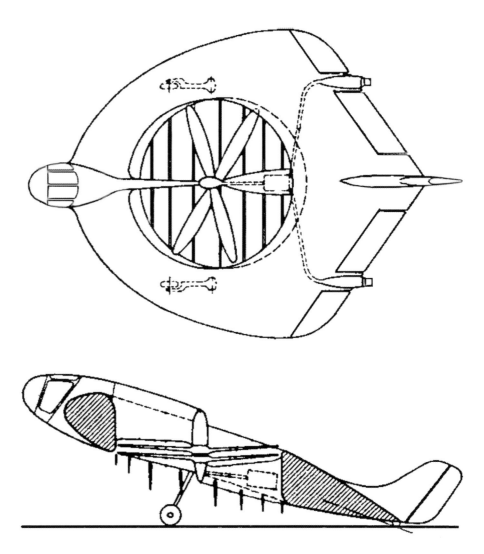

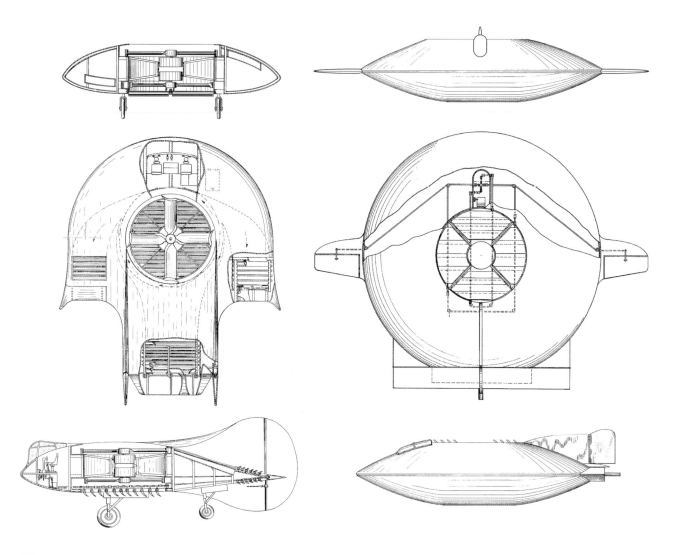

Above: **This VTOL design was produced by Charles Neumann and Hugo Baca in 1945. It shares some similarities with the Focke-Achgelis VTOL concept.**
US Patent Office

Above and below: **Although never built, Naught's 1948 design for a VTOL flying saucer using a ducted fan system proved influential with many designers.**
US Patent Office

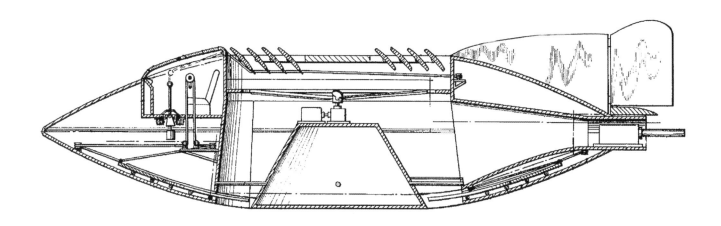

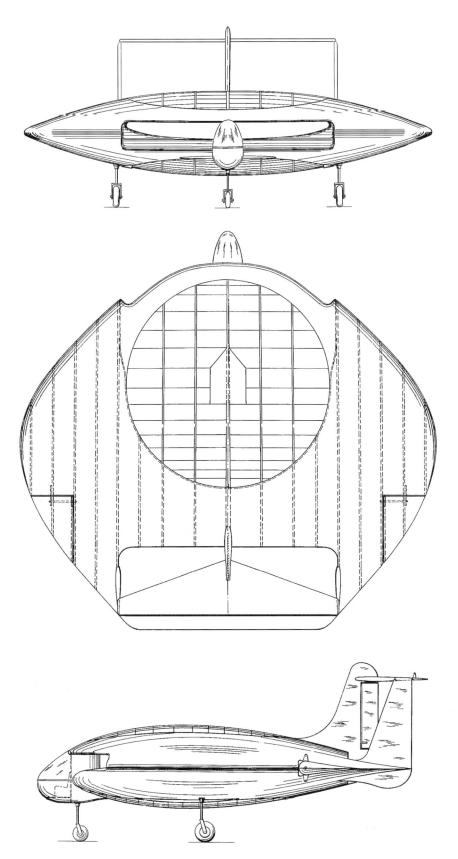

The Goodyear Convertible Aircraft designed by Robert Ross in the early 1950s was another proposal with similar objectives to the wartime Focke-Achgelis VTOL aircraft. *US Patent Office*

during World War 2 they became a major subcontractor for Vought, producing large numbers of F4U-1 Corsair naval fighters.

One month after Goodyear made a patent application for their Convertible Aircraft design, a Californian physician and aircraft enthusiast called Horner Frederick Streib (1915-2000) completed the first of four circular-winged aircraft studies. Each design was a variation on the same theme and they were all submitted to the US Patent Office during the 1950s.

As a starting point for his aircraft, Streib used a central duct containing two contra-rotating propellers with the powerplant above the blades. Two jet engines buried in the rear fuselage would be used for horizontal flight. As the flying disc evolved, Streib modified the central rotor assembly to tilt as required during VTOL operations. He then developed this idea further for one of his concepts, with the entire duct swivelling through 90° to provide propulsion for horizontal flight. Various options were considered for these concepts, such as jet engines in pods for horizontal flight and different arrangements for control surfaces and stabilising fins.

Streib also considered the possibility of boundary layer control and indicates that models of his designs were built and wind tunnel testing was undertaken. All these designs positioned the raised cockpit towards the leading edge and each aircraft would use a fully retractable tricycle undercarriage.

Exactly what role Streib envisaged for his design is unknown, although a military application seems possible. It is also worth noting that the earliest design was published in 1959, almost six years after it had been submitted. Reasons for the delay are unknown, but conflicts with classified projects remain a possibility. In 1955, the Focke-Wulf Company reappeared in Bremen as an aircraft manufacturer and the Board decided to dust off Professor Focke's Schnellflugzeug design and build a wooden scale model for wind tunnel testing.

At that time, the German military were very interested in VTOL aircraft capable of operating from improvised sites and the Schnellflugzeug design seemed to be worth a second look. Trials began in early 1957, with the aircraft now being called the Rochen (Aquatic Ray or Skate). But it soon became apparent that despite the VTOL capability, this aircraft would never compete effectively with jets. Consequently, the project was effectively dead by the start of 1958.

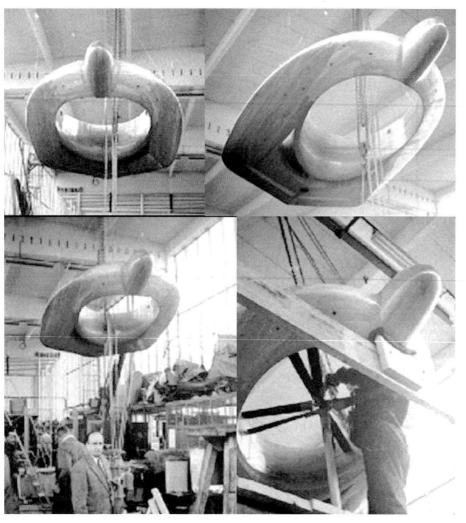

Above: **These images show post-war construction of a scale-sized Focke-Wulf VTOL aircraft design known as the Rochen (Ray).** *Bill Rose*

Left: **Designed by Homer Streib in the 1950s, this sophisticated disc-shaped aircraft was expected o provide a full VTOL capability and high-speed horizontal flight.** *US Patent Office*

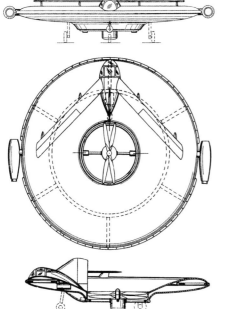

FOO FIGHTERS

From 1942 onwards, USAAF and RAF aircrews would occasionally file reports of strange (mostly ball-shaped) lights that appeared to shadow their aircraft across occupied Europe. This continued until 1944, when a substantial rise in the number of incidents seemed to suggest that a new form of German weapon was undergoing trials. As a consequence, immediate efforts were made to determine the threat.

US General William J. Donovan of the OSS (Office of Strategic Studies – a special ops CIA predecessor) was given the task of setting up a technical group to assess the situation. This would bring together some of America's best

scientists, such as Dr Howard P. Robertson, Dr David Griggs (1911-1974) and future Nobel Prize winner, Dr Luis Alvarez (1911-1988).

Aircrews were reporting that these mysterious balls of light would appear to operate under intelligent control. They could not be outmanoeuvred or shot down, but caused no obvious damage to the aircraft. So widespread were these nocturnal encounters that official censors allowed the Associated Press Corps in Paris to make the following rather vague release on 7 November 1944.

'The Germans are using jet and rocket propelled planes and various other "newfangled" gadgets against Allied night fighters,' Lt. Col. B. Johnson, Natchitoches, La., commander of a P-61 Black Widow group, said today. 'In recent nights we've counted 15 to 20 jet planes.' Johnson said. 'They sometimes fly in formations of four, but more often they fly alone.'

On 14 December 1944, The New York Times reported that floating silver balls seen during daylight hours were thought to be a new type of German weapon. Just over a month later, on 15 January 1945, Time magazine described these objects as 'the most puzzling secret weapon that Allied fighters have yet encountered'. The report suggested that they might be intended to dazzle pilots, act as aiming points for anti-aircraft gunners, interfere with airborne radar or interrupt the ignition system of an aircraft engine.

During the same week, Newsweek commented, 'Possibly they are the results of a new anti-radar device which the Germans have developed. On the other hand, they may be the exhaust trails of a smaller model of the radio-controlled Messerschmitt 163, a rocket-propelled flying wing. Day bombers have met the Me 163, which has an explosive charge in the nose and is apparently designed to crash into Allied planes. When one pilot closely inspected Foo Fighters tagging him, however, he detected nothing but the spheres.'

In fact, the Me 163 rocket-powered interceptor was never the cause. It was not designed to collide with Allied bombers and operated from only one base. The interceptor's range was extremely limited and it was incapable of flying at night. Nor can any sightings of these mysterious lights be linked to the brief use of Me 262 jet fighters adapted for night operations in the closing months of the war.

From November 1944 onwards, the term 'Kraut Fireballs', or more usually 'Foo Fighters', was in widespread use by American aircrews, with the latter being adopted by members of the RAF. The first formally documented use of the name 'Foo Fighter' took place in 1944 and is attributed to Lt Donald J. Meiers, who was a

Above: **Simulation of a Foo Fighter trailing RAF bombers across the war-torn skies of Europe. The most likely cause of these incidents is a still unexplained natural phenomenon.** *Bill Rose*

radar operator with the USAAF 415th Night Fighter Squadron.

The term 'Foo Fighter' had been taken from a syndicated US cartoon strip called 'Smokey Stover' that was created by Bill Holman (1903-1987). This series was carried by many American newspapers from 1935 until Holman's retirement in 1973. The central character Smokey Stover drove a fire truck called the Foomobile and described himself as a Foofighter. Apparently, Holman found the word 'Foo' on the base of a Chinese figurine. There were certainly a number of USAAF aircraft operating in Europe during 1944-45 that carried 'Smokey Stover' artwork and this may have initially encouraged aircrews to use the term.

When OSS General Donovan completed his report on Foo Fighters, it concluded that although the phenomenon was unusual, it was not new Nazi technology. The RAF also formed a small top-secret committee to examine the issue of Foo Fighters and, like the Americans, they decided that the cause was often some kind of unusual electrical phenomenon probably related to St Elmo's Fire.

There was no dispute about the authenticity of Foo Fighter reports describing mysterious aerial lights and floating silver balls, but the exact cause of this phenomenon remains unclear. Today, there are some UFO investigators with their own agendas who have convinced

themselves that these lights were of a supernatural nature, or perhaps small alien reconnaissance vehicles spying on mankind's aggressive activities. This particular idea found its way into the 1956 movie Earth vs. the Flying Saucers.

However, these suggestions aside, the idea that Foo Fighters were advanced robot aerial vehicles developed in great secrecy by Nazi scientists has gained some acceptance, despite the complete lack of supporting evidence. Nazi Germany was at the forefront of aviation development and this know-how was eagerly sought by East and West once hostilities ceased. The Germans had shown their ability to develop advanced air-launched guided weapons and were preparing to start production of surface-to-air missiles. With this in mind, is it so unreasonable to suggest that they may have been experimenting with small remote-controlled vehicles intended for development as anti-aircraft weapons?

Major Rudolph Lusar appears to have been unaware of any such development, but Renato Vesco (1924-1999) would later claim that Foo Fighters were advanced drones designed to emit radio frequency pulses that would interfere with the electrical systems of enemy aircraft. It seems that these machines were given the name Feuerball (Fireball).

Vesco was an Italian based in Genoa, who

claimed to have been an aircraft engineer, educated at the University of Rome prior to World War 2. After studying at a German Institute, Vesco worked at the huge underground Fiat facility by Lake Garda. According to Vesco, he was retained as a UFO investigator after the war by the Italian government.

Vesco produced several books on UFOs claiming that the Feuerball had been developed by a top-secret Luftwaffe research facility near Oberammergau in Bavaria. Apparently, engineers from Henschel and Zeppelin were recruited by the SS to develop this technology. It had a tortoiseshell-shaped body and was powered by a system of jets that spun around the vehicle's circumference, creating the 'ball of light' effect while in flight. The Feuerball could be equipped with technology designed to emit high frequency pulses that would interfere with equipment carried by Allied warplanes. A more advanced version of Feuerball called Kugelblitz (Ball Lightning) was said to be in development. According to Vesco, all the

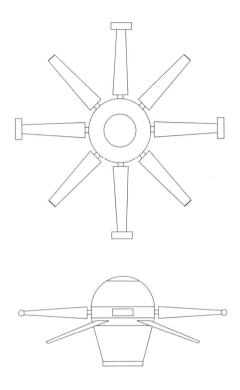

Left: **This is one of two configurations that have been suggested for a small robot vehicle used by specialist German units to interfere with Allied bombers during the later stages of World War 2. Generally referred to as a Foo Fighter by Allied aircrews and allegedly called the Feuerball (Fireball), this vehicle would have required present-day technology to make it viable and it is therefore – in the author's opinion – most unlikely to have existed. Another suggestion to explain Foo Fighters as German robot weapons describes a small disc-shaped craft powered by a miniature radial flow gas turbine.** *Bill Rose*

technical material relating to the Feuerball was retrieved by British Intelligence after Germany's surrender and immediately classified as top secret.

Vesco's story is certainly interesting and provides the kind of explanation for Foo Fighters that has won considerable support over the course of many decades. Unfortunately, there appears to be absolutely no truth to any of it. Vesco proved impossible to track down during the 1990s and it seems

that he was unwilling to speak with any journalists, researchers or writers. Much of the background information on Vesco, such as a section in the book Projekt UFO (1995) by William A. Harbinson, is incorrect. Furthermore, Vesco's own work is generally very hard to take seriously.

When hostilities began in Europe, Vesco would have been no more than fifteen years old, and he never attended the University of Rome nor any German Institutes. Nor could he have graduated as an aeronautical engineer. There is no evidence to show that he worked for Fiat at Lake Garda and he is believed to have undertaken military service during the war, operating in a clerical capacity at an office in Milan. In peacetime, Vesco is said to have worked on various merchant ships as a mechanic and finally retired in the 1960s, becoming something of a recluse.

This aside, the Allies were concerned about Foo Fighters and there is evidence that scientists believed the Germans might be developing technology that could interfere with the electrical systems of aircraft. Perhaps this particular concern reflected classified research already taking place in

Britain or America? In late 2002, the American aviation researcher Joel Carpenter discovered a previously secret USAAF Intelligence Log (UKX-23426) dated 27 January 1945. This interesting document warns of new German technology designed to interfere with the ignition systems of aircraft engines. It is suggested that component shielding may be necessary to counteract the effect. But the suggestion that Foo Fighters were Nazi weapons fails to explain the sightings of brilliant globes above the Pacific during the war with Japan. Conceivably the Feuerball technology was transferred to Japan, but there is absolutely no evidence to support this.

Reports also suggest the phenomena in the Far East was different, often taking the form of glowing balls of light that floated in the sky, with nothing to show that these objects interfered with any aircraft. Regardless of the true nature of Foo Fighters, it is very evident that there was genuine official concern about the increasingly frequent encounters with balls of light by Allied aircraft.

The real issue with this Nazi secret weapon hypothesis is a total lack of proof and the scientific limitations of that period. While it might be possible to build such a machine today, an advanced vehicle like the Feuerball was way beyond the capabilities of wartime German science, or for that matter any other country during the 1940s.

The propulsion system for such a vehicle might have just been feasible, but the type of guidance, sensors and control systems required were definitely not. An advanced flight management system providing some degree of autonomy simply belonged in the realm of SF, along with an electromagnetic pulse weapon designed to disable enemy aircraft. However, the widespread reports of glowing balls of light seen by Allied aircrews cannot be dismissed and are probably linked to some still unexplained natural phenomenon such as ball lightning.

At the present time there is nothing to convince this author that there is any substance to the many claims that Foo Fighters were small, very advanced, unmanned secret weapons. Everything seems to point towards this idea being simply the product of someone's imagination that has been endlessly repeated and given a life of its own.

THE SOVIET ARROW

The Moskalyev SAM-29 Strela (Arrow) was an astonishingly advanced proposal that almost did not make it into this book. That is because

One of the few existing photographs of the Moskalyev SAM-9 experimental aircraft built in the late 1930s. *Bill Rose*

it falls somewhere between a delta and a semi-elliptical shape. The rocket-powered Strela was based on a small, single-seat, piston-engined prototype known as the Moskalyev SAM-9 and developed in the mid-1930s by Professor Alexander Moskalyev (1904-1982). Variants of this design were proposed, with different engines, twin tail fins and differing dimensions.

SAM-9 was a compact aircraft, having a wingspan of 11ft 7in (3.55m), a wing area of 139.9sq ft (13m²) and a length of 20ft 2in (6.15m). Gross weight is quoted as 1,389lb (630kg) and the aircraft was built mainly from wood. It was supported on the ground by two fixed forward struts with wheels and there was a simple tail skid. The aircraft was powered by a Renault Bengali-6 (MV-6) in-line piston engine producing 220hp (164kW) driving a two-blade propeller. Performance figures for this aircraft are unclear, although SAM-9 is believed to have had a maximum speed of close to 200mph (325kph). Design work was completed in 1936 and the aircraft was assembled during the first half of 1937, followed by a protracted period of testing.

In 1944, when Moskalyev was invited to participate in the development of a high performance rocket-powered interceptor, possibly having the ability to reach Mach 1, he used SAM-9 as the starting point for this project. This elliptical-shaped interceptor known as SAM-29, or RM-1 for Raketnyi Moskalyev 1 (RM-1), would be powered by a Dushkin RD-2M-3V liquid fuel rocket engine producing an anticipated 4,400lb (2,000kg) thrust. Most of the metal airframe would be filled with fuel storage tanks and the aircraft was to be armed with two cannons.

A fully retractable tricycle undercarriage was planned and it is assumed that after an interception the aircraft would glide back to base. No dimensions or weights are available, but SAM-29 would have been somewhat larger than the SAM-9. Moskalyev received some technical assistance from the rocket pioneer Sergei Korolev (1906-1966), but ministry officials decided that SAM-29 was too advanced and cancelled the project when the war ended. Moskalyev's OKB was closed down in January 1946 and he accepted an academic position in Moscow. Apparently, he continued to work on this proposal until 1948.

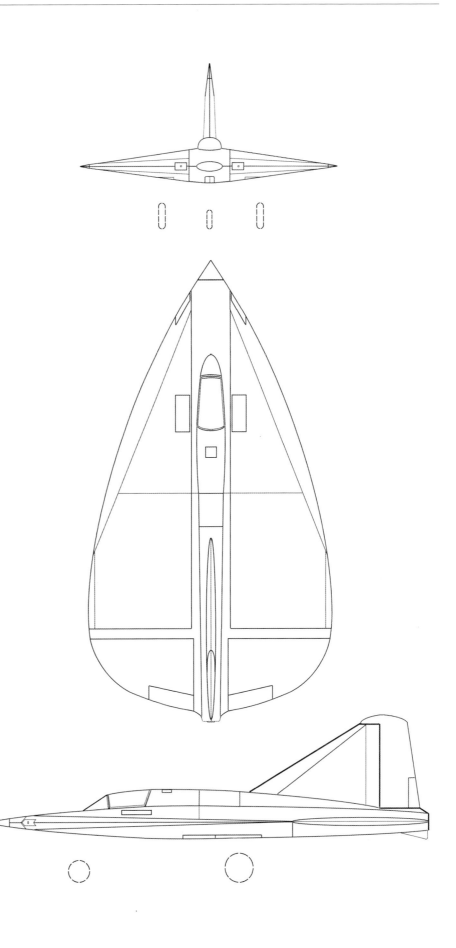

Right: **This drawing shows the anticipated appearance of the Moskalyev SAM-29 Strela (Arrow) rocket-powered supersonic interceptor. Designed during World War 2, this proposal was considered too far ahead of its time and abandoned.** *Bill Rose*

The Post-War Era

Following the Arnold incident of 24 June 1947, there was an explosion of UFO sightings across America and around the world. The term 'Flying Saucer' entered widespread use and the media took full advantage of the situation.

It began to look as if there was real substance to this new phenomenon, but the US and UK military insisted that it was mostly nonsense that was not worth bothering with. Nevertheless, behind the scenes there was genuine concern that something very unusual was taking place. If flying saucers were real, then there were two worrying possibilities. Either these craft were long-range Soviet spyplanes, almost certainly developed from captured Nazi technology, or they really were from another world, perhaps attracted here by our destructive activities or the early use of nuclear energy.

Although there have been vigorous denials, it appears that the RAND Corporation (Research ANd Development), which has acted as a think-tank for the CIA, began to study UFOs soon after Arnold's sighting. For reasons unknown, this work remains classified, although RAND is said to have considered various scenarios – from incursions of US airspace by advanced Soviet long-range aircraft to an extraterrestrial presence leading to the possibility of first contact, or an alien invasion. The CIA and USAF were already actively engaged in studying UFO phenomena and while CIA interest was entirely behind closed doors, the USAF were somewhat more open with their actions, having formally commenced an investigative programme called Project Sign, which was undertaken at Wright Field AFB (later Wright-Patterson AFB).

ALFRED LOEDDING, USAF PROJECT SIGN & DISCPLANES

Little known outside aviation and ufology circles, Alfred Christian Loedding (1906-1963) was a consultant engineer who became involved with many highly classified projects. He designed sophisticated aircraft, gained a deep knowledge of advanced German

Left: **Alfred Loedding with flying saucer model at Wright-Patterson AFB.** *USAF*

wartime aviation technology and played a key role in official UFO investigations during the late 1940s.

Loedding held the highest security clearance and had access to information that in some cases remains secret. He worked for USAAF Intelligence (sometimes in conjunction with the FBI or CIA) and, even today, Loedding's name is missing from US Social Security records, perhaps reflecting the secretive nature of his activities.

Shortly after graduating from the Daniel Guggenheim School of Aeronautics at New York University in 1930, Loedding filed his first US patent (1897669) for a small and rather neat monoplane with an unusual supplementary airfoil. He then secured employment with the US War Department, working as an engineer in various capacities until 1937, when he joined the Bellanca Aircraft Co of Delaware as manager of their Stress Analysis Department. In 1938, Loedding accepted a civilian job with the US Army Air Corps (USAAC), working for various

Right: **General Nathan Twining, Commander USAF Air Material Command, who formally suggested the formation of a specialist unit within T-2 to gather information on unidentified flying discs. The new department was approved in late 1947 and became Project Sign.** *USAF*

technical departments at Wright Field, Ohio.

During World War 2, Loedding acted as a consultant and advisor to the Technical Intelligence Division of the Air Material Command (AMC), known as T-2. He was involved in setting up the original Jet Propulsion Laboratory (at Wright Field) and became an expert on wind tunnel design, aerodynamics and rocket propulsion.

Loedding's family originated in Germany and he was fluent in the language. This became very useful at the end of the war when Germany's leading aeronautical scientists began to arrive at Wright Field for debriefing. Loedding made friends with many of these former enemies, often inviting them to his house for dinner. He was highly respected by his colleagues and developed close personal friendships with a number of leading scientists such as Robert Goddard, Theodore von Karman and Alexander Lippisch.

At the end of the war, Loedding and von Karman were deeply involved with the evaluation of captured German aviation technology, which would often be passed to appropriate defence contractors for further

Flying Saucer Technology

Above: **Walter (left) and Reimar Horten (right) worked at the forefront of German wartime flying wing development. In the immediate post-war years, US Intelligence considered them to be behind the secret development of flying saucers, perhaps operated by the Soviets.** *David Myhra*

study. This continued until summer 1947 when T-2 felt their task of collecting and examining wartime German and Japanese aircraft and equipment had run its course.

On 24 June 1947, Kenneth Arnold made his historic sighting and this was followed by an unexpected wave of UFO sightings that showed no signs of abating. As a consequence, Technical Intelligence at Wright Field began to take an informal interest in UFOs, with Brigadier General George Schulgen and Alfred Loedding jointly examining this new phenomenon in some detail. Within weeks, Schulgen completed a brief review, which concluded that flying discs were almost certainly real craft. He then approached General Nathan Twining, who was in overall command of the Air Technical Service Command, suggesting that this issue required urgent attention. Twining accepted the recommendations and in January 1948 he assigned Captain Robert R. Sneider to head an investigation called Project Sign (initially Project Saucer) as a unit within the T-2 Technical Analysis Section.

Loedding had been discussing just such an idea with Twining and as a consequence he was asked to become the senior civilian member of the team. The Project Sign group was relatively small and included another civilian engineer called Lawrence Truettner, who specialised in missile technology, and Albert Deyarmond, a civilian intelligence analyst. In addition, a handful of mainly civilian administrative personnel were attached to the unit on a part-time basis.

Loedding and Schulgen both appear to have been in agreement that the most likely explanation was long-range Soviet reconnaissance aircraft developed from advanced Nazi technology. They had talked with many German scientists and examined captured hardware such as the largely completed and very advanced Horten flying wing jet fighter. They also studied many of the German proposals for future aircraft and it is conceivable that Loedding interviewed Richard Miethe and knew about the Prague project.

Having given the matter considerable thought, Loedding finally concluded that any currently flying discs were probably based on designs produced by the Horten Brothers. He was aware that the Parabola glider had been damaged after construction, never flew and was finally set on fire and destroyed. But Loedding believed that their crescent-shaped Parabola was just the starting point for more advanced designs, somewhat larger in size and perhaps powered by some newly devised form of nuclear propulsion. One document drafted by Loedding and Schulgen which discusses this suggests that 'atomic energy engines would probably be unlike any familiar type of engine'.

A declassified USAF document from General George Schulgen that bears his name, dated 30 October 1947, discusses UFOs in some detail and Schulgen indicates that he thought the Hortens might be responsible for designing some of the unidentified flying discs seen around the world. Schulgen said, "What is the present activity of the Horten brothers, Walter and Reimar? What is known of the whereabouts of the entire Horten family?" Then he asks, "Has there been any interest shown by the Russians to develop their aircraft? Are any efforts being made to develop the Horten 'Parabola' or modify this configuration to approximate an oval or disc?"

The Horten Brothers were now central to the preliminary USAF investigation into flying discs. Nobody seemed to know their whereabouts and some suspected they were behind the Iron Curtain. These enquiries started in autumn 1947 and would continue for some months after Project Sign had been initiated.

Both Reimar Horten (1915-1993) and his brother Walter (1913-1998) had begun to

Below: **The experimental Horten Ho-9 V1 flying wing being moved in preparation for a test flight.** *Bill Rose*

design and build sophisticated flying wing gliders in Germany during the early 1930s. With the war in Europe under way, their work was officially recognised and they were given facilities and resources to develop advanced jet-powered flying wing fighters and bombers. When hostilities ceased, the first batch of prototype jet fighters were being assembled at the Gotha factory near Friedrichsroda.

USAF officials continued to make enquiries about the Hortens, unaware that they had been working for the British and were both in Germany. Walter Horten actually wrote to Jack Northrop in 1947, seeking employment with his aircraft company in California. In response, Northrop immediately offered design jobs to both the brothers, telling Walter to contact the USAFE in Wiesbaden and make formal applications to work in America. But his request was turned down, which remains perplexing, bearing in mind that significant numbers of USAAF personnel and various contractors were engaged in analysing the Hortens' wartime research.

Perhaps there was some underlying reason for this rejection, but it now looks as if there was a complete breakdown in communication between different US agencies. The 970th Counter Intelligence Command finally managed to locate the Horten Brothers in March 1948, and having established they were British assets, approval to make contact was sought by the British Joint Services Mission in Washington DC. Once this was arranged and interviews were undertaken, the records show that the Americans had 'collected all the desired information' that was required to conclude these enquiries.

It later transpired that the Hortens had considered building a powered successor to the Parabola, which might have eventually resulted in a high performance fighter. Further evidence for this line of development came to light in 2004, when several previously classified USAF documents from 1948 were located, which appear to show a sketch by one of the Hortens for an advanced horseshoe-shaped rocket-powered fighter.

This aircraft may have been expected to achieve supersonic speed at high altitude and would be powered by two rocket engines at each wingtip. There was a central semi-capsule-shaped cockpit containing the pilot in a prone position, and armament has been suggested as two cannons. It is not known how the aircraft would take off, although a rocket-powered trolley is a possibility and a central skid would be deployed for landing.

Some very rough calculations indicate the aircraft's overall length being about 68ft 10in (21.0m), with a span of 46ft (14.0m). The wing shape is very unusual, with a very thin leading edge increasing in width to a thicker section towards the trailing edge. A boundary layer control system was envisaged, although this strange design would have presented many aerodynamic problems.

Doubts have been raised about the authenticity of this document, which is marked MOIC Headquarters, CIC Region 1, File 1-1606 and mentions Operation HARASS. Conceivably this drawing was an attempt by one of the Hortens to show USAF Intelligence officials the type of experimental aircraft that might have been responsible for the Kenneth Arnold UFO sighting on 24 June 1947. Although the Hortens worked on some of Germany's most advanced wartime aviation projects, there is no evidence that they were ever involved with the design or development of flying discs. (For further details of the Horten Brothers' wartime projects see *Flying*

Top: **The completed Horten Parabola that led US Intelligence to believe that the brothers had been engaged in flying disc development.** *David Myhra*

Above: **The Horten Parabola glider was intentionally destroyed at Gottingen after it was decided that the aircraft was beyond repair.** *David Myhra*

Wings and Tailless Aircraft by Bill Rose, 2010.)

While the Hortens had now been eliminated as post-war flying saucer designers, Loedding and his colleagues continued to believe that the unidentified discs were real 'nuts and bolts' machines and the Soviets were the only people capable of building and operating them. Naturally, this opinion remained behind closed doors and there was serious debate about whether the USAF should set up observation posts and try to intercept flying discs with jet fighters. That said, the official position remained unchanged and the USAF maintained that almost every UFO report had some sort of explanation that ranged from simple misidentification to public hysteria and hoaxing.

During the months following the formal establishment of Project Sign, the USAF team investigated all the high-profile incidents from Arnold's sighting to Roswell and the Mantell Incident. As UFO sightings continued to increase, Sneider and Loedding slowly changed direction and started to think the unthinkable. Was it possible they were dealing with extraterrestrial vehicles after all? In late 1948, Sneider and Loedding completed a large document called 'Estimate of the Situation' that favoured an extraterrestrial presence on Earth.

It was immediately classified as top secret and presented to Colonel Howard M. McCoy, who forwarded copies to the Pentagon. There are allegedly references to an earlier, still unacknowledged, RAND UFO study and although Washington rejected Project Sign's findings, this document has never been released to the public. Consequently, Project Sign was disbanded and immediately replaced by a new UFO investigation group called Project Grudge, headed by Captain Edward J. Ruppelt. Project Grudge found no evidence of unknown aircraft of foreign origin operating within US airspace, and their office was closed on 27 December 1949.

Project Grudge was eventually followed by Project Blue Book, which seems to have concentrated on public relations issues

rather than serious investigation. For many years, Blue Book gathered data on UFO sightings that proved useful for analysing public attitudes. It is unclear if they set out to deliberately debunk sightings, but this may have occurred if a report from the public revealed a classified project or something unexplained with worrying implications. The Air Force certainly seemed happy with the way things were handled and the Blue Book office remained in operation until the end of 1969. Following Project Sign's closure, Loedding was reassigned to an engineering administration job in another section.

Aside from being a key player in the USAF's initial UFO investigation, Loedding had been designing aircraft since he left college and some of them were surprisingly advanced with many interesting features. Of particular note was a futuristic (although seemingly underpowered) single-engined, propeller-driven flying wing aircraft, which he filed as a patent on 18 May 1933. This was published on 24 May 1938 (2118254) and Loedding continued to evolve the design throughout World War 2.

After he became involved with UFO investigation and Project Sign, Loedding became interested in the idea of designing a flying saucer, considering it a viable proposition. He built several small desktop-sized models that were oval or completely circular and these may have been wind tunnel tested. This led to the design of a very advanced semicircular flying saucer concept, which was filed as a patent (2619302) on 25 August 1948. It is interesting to note the design for this craft would not be published for another four years.

Loedding's low aspect ratio, four-seat utility aircraft was oval-shaped in appearance with a flush, substantially glazed forward cockpit and no external protrusions apart from two angled tail fins with control surfaces and two small rudders. Loedding appears to have subsequently replaced this arrangement with a single, large, upright stabilising fin and rudder, plus two larger control fins at the trailing edge. For propulsion, the aircraft would use a ducted fan arrangement, drawing air through louvred panels and expelling it via a

Right: **An advanced pre-war aircraft design produced by Alfred Loedding. It seems probable that the propulsion system as envisaged would have been inadequate.** *US Patent Office*

Below: **This drawing of a supersonic rocket-powered interceptor is said to have been produced by one of the Horten Brothers in the late 1940s and appeared in an official US military document.** *Bill Rose*

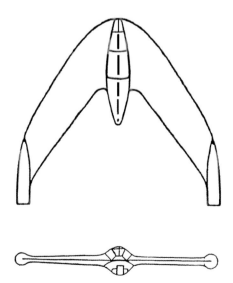

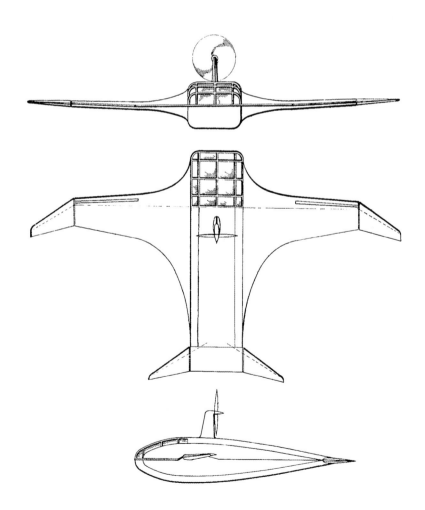

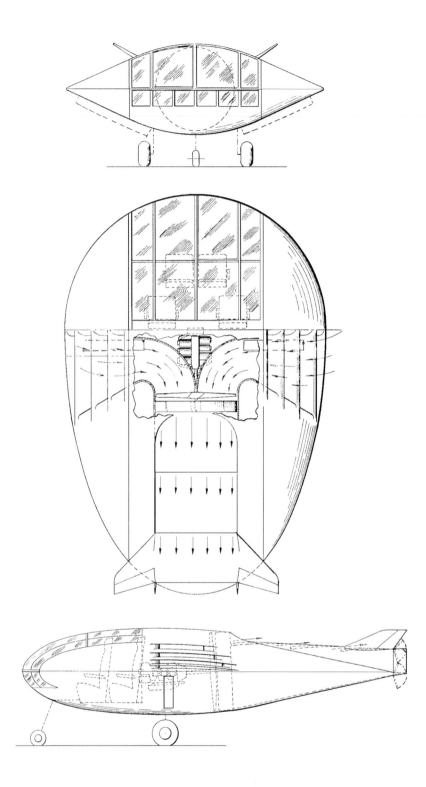

Right: **Loedding's advanced design for an oval-shaped VTOL utility aircraft using a ducted fan propulsion system.** *Bill Rose, based on a US Patent drawing*

There were also features such as opening flaps near the air inlets to allow a controlled descent in the event of an engine failure, which could be used as air brakes during a normal horizontal landing. On the ground, this aircraft would be supported by a fully retractable tricycle undercarriage. What makes this concept particularly interesting is the fact that Loedding was one of America's leading experts on wartime German aviation technology. Some of those advanced projects undoubtedly influenced his futuristic concept.

Whether or not any part of this proposal was connected to the supposed Heinkel-BMW project remains unknown, although Loedding's colleague Dr Theodore von Kardan gave top priority to investigating every aspect of BMW's wartime research after hostilities ceased. The design for Loedding's aircraft was never taken up, but it proved to be influential with many leading aeronautical designers and engineers. These included John Frost, who devised the Avrocar; the legendary Alexander Lippisch; Nathan Price of Lockheed, who worked on a secret flying disc project; and A. A. Griffith of Rolls-Royce.

From all accounts, Loedding was a brilliant scientist, although he does appear to have lost his objectivity while dealing with the issue of UFO phenomena. He was certainly not alone in this respect. Loedding finally resigned from his civilian job with the USAF in 1951. He then became Director of Jet Research for the Unexcelled Chemical Corporation at Cranbury, New Jersey, overseeing R&D on missile propellants and filing a number of patents for flares, rockets and fuel formulations. In 1955, he rejoined the USAF at Wright-Patterson AFB and was eventually appointed the Civilian Liaison Officer between the USAF and NASA. On 10 October 1963, Loedding died in Williamsburg, Virginia, from cancer.

LIPPISCH HEEL-SHAPED AIRCRAFT

German-born Alexander Martin Lippisch (1894-1976) was arguably one of the greatest aeronautical designers of the last century. He worked for a number of well-known aviation companies before starting to develop gliders. By the mid-1920s, Lippisch had become Director of the Aeronautical Department of the Rhön-Rossitten Gesellschaft (RRG) that

system of slats on the upper rear fuselage surface of the tail fins. It is less clear how this would work on the single-fin version, but it is probable that these features were moved to the underside.

The powerplant envisaged for this design was either a single water-cooled internal combustion engine driving a fan, or a gas turbine. Loedding was interested in the possible use of boundary layer control and while he did not suggest any performance figures for this aircraft, he envisaged a high-speed capability. He also anticipated very good STOL performance and perhaps full VTOL, presumably dependent on adequate engine power.

Above: **The legendary German aerodynamicist Alexander Lippisch, who worked on many advanced wartime aircraft projects.** *Bill Rose*

Right: **An unusual post-war elliptical ducted fan aircraft design produced by Alexander Lippisch.** *US Patent Office*

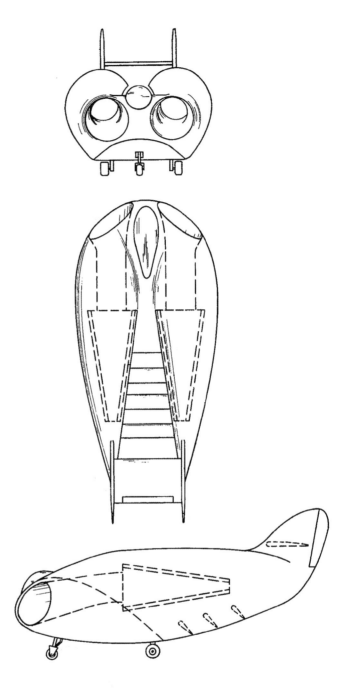

later became the German Research Institute for Soaring Flight (Deutsche Forschungsanstalt fur Segelflug – DFS).

He then began work on his first powered aircraft called 'Storch' (Stork) and was responsible for early development of the delta wing, coining the descriptive term for this shape from the triangular Greek letter. In the late 1930s, Lippisch gave up his position with the DFS to join Messerschmitt, where he developed the world's first operational rocket fighter. During the remainder or the war, Dr Lippisch produced many designs for highly advanced jet- and rocket-powered aircraft. He acted as a consultant for several companies, and when the war ended he was tracked down by US Intelligence and immediately recruited to work in America.

After completing a period of service for the US military, Lippisch joined the Collins Radio Company. He continued to undertake aviation design work and turned his attention to something new that drew on the principles used by Loedding for his low aspect ratio aircraft. Lippisch called this design a 'Fluid Sustained and Propelled Aircraft' and its overall appearance concept was more beetle-shaped than Loedding's. There would be two tail fins with rudders joined by a horizontal stabiliser with a control surface, and the forward-placed cockpit would provide excellent visibility for the crew. The aircraft would be supported on the ground by a conventional fully retractable tricycle undercarriage.

Propulsion took the form of two ducted contra-rotating rotors, driven by one or more unspecified engines that would probably be gas turbines. Two large forward-facing inlets would duct air to the rotors and it would then be deflected rearwards for horizontal flight or downwards via several louvred slats during VTOL operations or hover. One of the aims of this wingless design was to significantly reduce drag and allow a high forward speed, but no details of these expectations were published. Nor are there any references showing anticipated dimensions or weights.

The design was completed in 1954 and filed as a patent (2752109) on 6 October 1954. It is unclear if Lippisch intended this concept to be anything more than a general purpose utility aircraft, but it never progressed any further. It is probably true to say that this was the closest Dr Lippisch came to any kind of flying saucer design. Throughout the years, he produced numerous, sometimes rather impractical, concepts that ranged from supersonic fighters to annular wing aerodynes. He worked at the cutting edge of aeronautical design and, even today, his signature can be seen on many of the latest aircraft.

FICTIONAL FOO FIGHTERS

By the late 1940s, the idea of alien invaders visiting our planet was entrenched in the public's mind. Hollywood was preparing to capitalise on this with big budget productions such as *War of the Worlds* (1953) and *The Day the Earth Stood Still* (1951), but a now almost forgotten 15-part cinema serial is generally acknowledged to have included the first cinematic portrayal of a man-made flying saucer. This budget production from 1949 called *Bruce Gentry – Daredevil of the Skies* was developed from a New York Post Syndicate comic strip.

Produced for Columbia by Sam Katzman, the series was essentially a run-of-the-mill action adventure with hero Bruce Gentry (Tom Neil) pursuing an evil mastermind called The Recorder. As the plot unfolds we learn that The Recorder has developed unmanned flying saucer technology which can be made to behave like a missile or carry equipment that interferes with the electrical systems of aircraft. Unfortunately, the production lacked any kind of scientific credibility and at the end of the serial, one of The Recorder's henchmen launches a flying disc against the Panama Canal.

Despite the ridiculous plot and third-rate special effects, this series successfully utilised wartime reports of Foo Fighters and SF concepts to produce something new, combining the world of fiction with perceived reality.

STASINOS DISCPLANE

An interesting case of a flying saucer concept becoming something it was not can be traced back to 11 November 1950. This was the date when Nick Stasinos, an engineering graduate from the Northrop Aeronautical Institute, showed members of the media his design for a jet-powered single-seat discplane.

This highly detailed scale-sized model was reportedly 2ft 6in (762mm) in diameter and represented a full-sized circular-shaped aircraft that would be 25ft (7.62m) across, with a centrally positioned cockpit and a bubble canopy. On the ground a fully retractable tricycle undercarriage would support the aircraft.

The propulsion system was relatively complex, with two internal jet engines on each side of the cockpit for horizontal flight. The outer section of the aircraft was a rotating ring used to generate lift and powered by a further eight turbojets. Each of these engines would channel exhaust gas through the rotor. This system has some similarities to a later design produced by the French aeronautical pioneer René Couzinet.

How Stasinos intended to control the disc

in flight remains unclear, as there are no visible control surfaces (on the upper surface). It might have been possible to vary the output of the main engines during horizontal flight, but there are no other obvious methods. The exact role envisaged for this design remains unclear, although Stasinos may have simply envisaged it as an experimental aircraft.

For more than half a century, a photograph of the model has freely circulated within the UFO community. It has often been described as either a mock-up built for an exhibition, or a secret experimental Northrop flying saucer fighter waiting on the runway. The letters and numbers on the fuselage have also generated considerable interest, although it seems clear that NS is just the designer's initials.

It is said that Stasinos passed his flying saucer model to Ripley's Believe It or Not museum in New York. However, Ripley Entertainment's senior archivist, Edward Meyer, advised me in 2005 that the model was not kept at any of the company's museums and no member of staff could ever recall having seen it.

DOAK VTOL CONCEPT

An interesting concept for a semi-saucer-shaped aircraft was produced by Edmond Rufus Doak (1898-1986) in the late 1940s. Doak was born in Texas, but his family relocated to California where he worked for a string of local aircraft companies that included the Glenn Martin Aviation Company, North American Aviation and Douglas.

In 1939, he gave up his job as Vice President of Douglas-El Segundo to establish the Doak Aircraft Co at Torrance, which is just south of Los Angeles. Doak's company built plywood trainers until the US entered World War 2 and then became a subcontractor for

Vultee, North American and a manufacturer of smaller components for most of the well-known Californian aviation companies. By the end of the war, Doak employed almost 4,000 women on war production and it was then necessary to downsize the business and diversify into new markets.

Nevertheless, the company continued to manufacture specialist aircraft components and Edmond Doak began to take a keen personal interest in helicopters and VTOL concepts, with hopes of developing new products that would revolutionise the aircraft industry. By the late 1940s, he appears to have secured USAF interest in the design of a VTOL aircraft that would use a ducted dual fan propulsion system for lift and forward flight.

His concept was not fully circular, with a straight trailing edge fitted with two elevons. On the upper rear of the fuselage was a stabilising fin, and the documentation indicates that this would be fitted with a rudder. Small wingtip fins would provide additional control.

The cockpit was positioned towards the leading edge, with room for two side-by-side occupants and directly behind was the duct containing the propulsive system. Drawings show the duct angled at 23°, but this might have been greater in a prototype. The duct contained two coaxial contra-rotating sets of rotors, configured to eliminate torque and driven by an unspecified engine in the aft fuselage section. It seems probable that Doak intended this to be a gas turbine. Air would be directed downwards to provide lift and it would be channelled rearward for horizontal flight using hydraulically controlled vanes that were positioned in the airflow.

An initial design by Edmond Doak for a semi-circular ducted fan VTOL Aircraft: *US Patent Office*

An elliptical aircraft design produced by Edmond Doak in the late 1950s, utilising four swivelling ducted fan units for lift and horizontal flight. *US Patent Office*

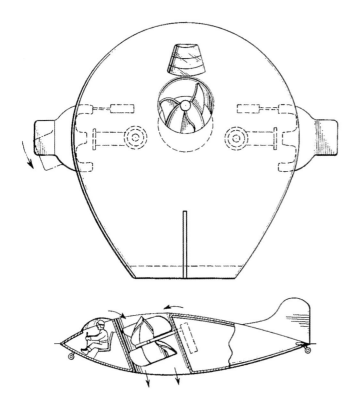

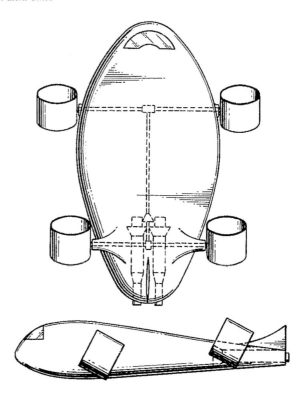

Doak also considered a fully retractable undercarriage for the aircraft. A very rough estimate suggests the size of the proposed vehicle would have been about 26-28ft (7.9-8.5m) in diameter. Unfortunately the details of this concept are rather vague and it appears that Doak was more interested in developing the propulsion system than the aircraft. This was certainly tested in model form and the design was filed as a US patent in 1950, with publication almost six years later (2730311). Nothing came of Doak's flying saucer design, but this led to a series of concepts for aircraft equipped with swivelling ducted fans that would have superior performance to a helicopter.

On 10 April 1956, the US Army issued a development contract for a proof-of-concept prototype designated VZ-4DA, serial number 56-9642, that was extensively tested by the US Army and NASA, although the performance was judged insufficient for further development. Doak Aviation ceased trading in 1961 and rights to the VZ-4DA were sold to Douglas Aviation in Long Beach, who never exploited the concept further.

EXTREME MANOEUVRES

The possibility of building a manned flying saucer capable of very sharp turns that were impossible with a conventional aircraft was apparent by the early 1950s. The first engineer and designer to fully develop this idea was John C. Fischer Jr who completed a set of proposals for a high performance flying disc in 1954. At this time, Fischer appears to have been working for the Ryan Aeronautical Co at San Diego, California, but it is unclear whether Ryan sponsored this unusual project or Fischer developed it independently.

He proposed a lenticular-shaped aircraft with a lower stabilising fin and rudder and an engine air intake beneath the leading edge. The overall diameter of this flying disc was an estimated 50ft (15.2m). A single-seat cockpit was centrally located on the upper surface of the aircraft and covered by a dome-shaped transparency. Two unspecified turbojet engines buried within the upper fuselage would provide propulsion during horizontal flight. A VTOL capability was not considered and the disc would operate from a normal runway.

In flight, the aircraft would be controlled by two large elevons along the trailing edge and a rudder fitted to the lower stabilising fin.

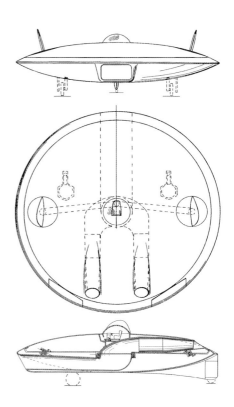

John C. Fischer Jr designed this flying disc in 1954 as a highly manoeuvrable fighter. *US Patent Office*

The most important hidden feature of this design was the swivelling upper shell of the aircraft that could be rotated by the pilot. This would be achieved by the use of two upright fins, with the option of power assistance at high altitudes. It was thought that relatively high g manoeuvres would be tolerable because of the pilot's fixed location at the centre of the aircraft, and fast turns could be made without banking.

The upper shell would rotate on roller bearings and it would be braked in a forward-facing position during normal horizontal flight. The aircraft was supported on the ground by a fully retractable undercarriage comprising two forward struts with wheels, plus a tail wheel carried within the stabilising strut.

One major problem with this and many other flying disc designs was pilot visibility, and while the idea of a high-speed disc-shaped fighter with extreme manoeuvrability is interesting, visibility remains a major issue with centrally located cockpits. Fischer's design failed to attract any official sponsorship, but the use of a flying disc for high g manoeuvres was later proven by the experimental Pye Wacket air-to-air missile system.

DAVIS FLYING SAUCER

There are numerous flying saucer patents on file, ranging from highly credible designs produced by teams of engineers working for aviation contractors to really weird contraptions that have more in common with SF. Some independent designs like the Weygers Discopter have proved incredibly influential and the patent for this proposal is cited in numerous technical documents from the late 1940s to the present day. One particularly interesting design influenced by the Weygers Discopter was produced by a Texas-based inventor called John W. Davis in the mid-1950s.

Davis proposed a circular-winged aircraft utilising an enclosed rotor system consisting of two coaxial contra-rotating assemblies configured to eliminate torque problems. Each set of four rotor blades would revolve around a central hub and they would be attached to two separate turbine rings that were driven in opposite directions by the exhaust from a turbojet engine located directly below the central hub.

Air for the engine would be drawn in from a forward ventral intake and the exhaust leaving the turbine blades would pass across a rudder and elevator assembly. In addition, there would be wingtip ailerons providing control and the ability to regulate the speed of the rotors. Full VTOL and hover would be

possible, plus high horizontal speed, when a series of shutters would be used to partially close the rotor housing. The aircraft would have a single centrally located cockpit covered with a blister canopy and it would be supported on the ground by a fully retractable tricycle undercarriage.

Davis claimed that his design would be relatively inexpensive to build and efficient in operation, although the development of this design would have presented a number of engineering challenges. He intended his design to fulfil some kind of military role, but although a versatile concept, it probably lacked sufficient purpose to warrant further development.

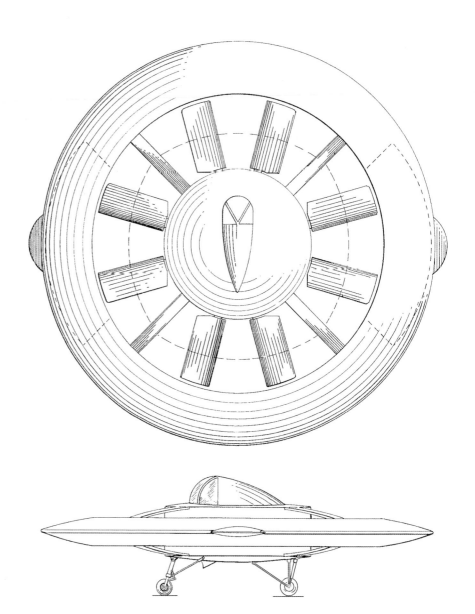

An advanced gas turbine-powered rotor design produced by John W. Davis. *Bill Rose*

THE NEW YORK UTILITY DESIGNS

The design for a small circular-shaped aircraft was completed in November 1953 by Leroy W. Crookes (1920-2005), who produced several engineering innovations and may have worked as a Patent Agent. It was intended to be an affordable utility aircraft, drawing on the earlier work of Johnson and Hoffman. Crookes anticipated excellent STOL performance, good stability and structural strength.

His proposal was an almost completely circular shape with a diameter of 18ft (5.4m) and propelled by a forward-positioned internal combustion engine with a proposed rating of 65-125 hp (48-93kW). This would be

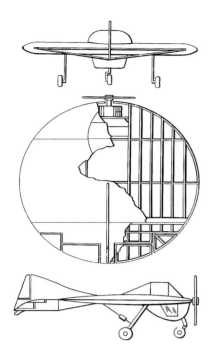

Left: **This small circular utility plane was designed in the early 1950s by Leroy Crookes but never reached the construction phase.** *Bill Rose*

directly coupled to a tractor propeller. As a future alternative, Crookes envisaged a centrally located gas turbine with two air intakes at the leading edge and an exhaust duct extending to the centre of the trailing edge.

The aircraft would carry 2-4 people in a forward-located cockpit, which required a foot well that protruded below the fuselage.

Left: **This small circular utility plane was designed in the early 1950s by Leroy Crookes but never reached the construction phase.** *Bill Rose*

This would be fitted with transparent sections to provide adequate downward visibility. The aircraft would be supported on the ground by a fixed position tricycle undercarriage and controlled in flight by ailerons located on the trailing edge and a rudder attached to a vertical stabilising fin. It was planned to build most of the aircraft from aluminium sections and tubing, with a continuous tube being used to form the aircraft's leading edge.

Crookes suggested that his basic design would be suitable for a range of aircraft, from small models to large transporters. What makes this concept particularly interesting is the way it is directly linked to another design completed about three months later in 1954 by another designer based in New York State called Curtis Donald Kissinger.

He produced basic plans for an affordable, compact, circular-winged aircraft with exceptional STOL performance and improved stability over conventional designs. In appearance, the wing was almost circular with an 18ft (5.4m) diameter and dihedral tips. There were two upright fins with rudders and a horizontal stabiliser located between them, fitted with an elevon. In addition, two small triangular ailerons were positioned behind the dihedral sections. The cockpit

was well positioned at the front of the aircraft, and there would be accommodation for at least one additional person and perhaps two more.

Kissinger proposed the use of a centrally located internal combustion engine providing 65-85hp (48-63kW) that would be shaft coupled to a forward two-blade propeller. A fixed undercarriage would support the aircraft on the ground, consisting of two forward struts with main wheels and two smaller tail wheels. It was hoped that this concept might form the basis of a family of larger aircraft carrying a number of passengers or cargo within the wing. These aircraft might be powered by turbojet and would have fully retractable undercarriages.

Although there are various differences between the Crookes and Kissinger aircraft, they are both exactly the same size, have similar performance expectations and were patented within the space of a few months. Furthermore, reading the technical details, it becomes clear that the text was written by the same person, with continual duplication of detail and phrasing. At the present time, the connection between the two men and the designs remains unclear.

The Crookes design was published as a US patent on 10 January 1956, and Kissinger's aircraft patent was published on 16 December 1958. These aircraft proposals fell somewhere between the Arup S-2 and the Sack AS-6 V-1 in dimensions, engine performance and appearance. Little if any interest appears to have been shown in either design, although it is possible that Crookes or Kissinger built and tested a scale model or commissioned wind tunnel trials. Later documentation produced by Kissinger accepted there were shortcomings with his design and he began a major revision of his flying saucer aircraft during the late 1960s.

He now produced plans for a larger circular-winged aircraft with a distinct fuselage section and forward cockpit. Supported on a fixed tricycle undercarriage, the aircraft was propelled by an internal combustion engine in a nacelle near to the centre of the wing. This would drive a rear-mounted pusher propeller. Consideration was also given to the possibility of using a jet engine. There would be twin tail fins connected via a horizontal stabiliser, which

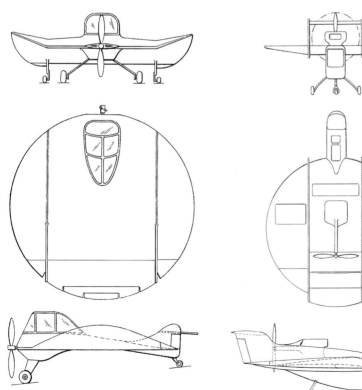

Far left: **Kissinger's original design for a disc-shaped light aircraft.** *US Patent Office*

Left: **Kissinger's later design for a circular-winged utility aircraft, which bears no obvious relationship to his earlier concept.** *US Patent Office*

was an elevon. Additional flight control would be achieved using ailerons, rudders mounted on the tail fins and several spoilers.

Kissinger believed that his new design would provide lower take-off and landing speeds as well as improved stability and better manoeuvrability. The US patent for this design was released on 18 March 1975, and although a considerable improvement over the early 1950s proposal, it failed to attract any commercial interest.

CANADA'S FLYING SAUCERS

As the Cold War intensified during the late 1940s, a major East-West conflict seemed increasingly likely. The Soviet Union was pushing ahead with the development of nuclear weapons, but it would be years before it possessed effective strategic missiles or the ability to manufacture compact nuclear warheads for them. This meant that any attack on Western Europe, Canada and North America would be undertaken with long-range bombers.

In the immediate post-war years, the most advanced bomber fielded by Russia was the Tupolev Tu-4 Bull, an unauthorised copy of the Boeing B-29. With improving Western defences, this represented a limited threat. Nevertheless, the CIA was confidently predicting that the Russians had started building fleets of jet-powered flying wings based on wartime Horten designs. This never came about and appears to have been alarmist speculation, but the reality was actually worse with the Soviets working on new long-range bombers able to match the most advanced American designs.

By 1952, the Tupolev Tu-95 Bear and the Myasishchyev M-4 Bison had both reached the prototype stage and when this became known to Western analysts there were very real concerns. It was also obvious that any threat to North America would come via the North Pole and mounting an effective defence would be difficult and extremely costly.

The inadequate American northern radar network utilised primitive, often unreliable World War 2 technology and it was appropriately called the 'Lashup' system. There were long-standing plans for a more advanced chain of radar stations, but this programme would not begin until 24 February 1954 when President Eisenhower approved construction of the newly named Distant Early Warning (DEW) Line.

There was also a lack of appropriate defensive weaponry, such as suitable combat aircraft. Subsonic all-weather fighters were

Above: **Two Avro-Canada CF-100 all-weather fighters. John Frost was recruited to manage the CF-100 programme in 1947.** *RCAF*

beginning to enter USAF and RCAF service, but it was feared that by the time sufficient numbers became available, the Russians would be operating more advanced bombers. RAF intelligence was in total agreement with the USAF and suggested that by 1960, the Russians would be flying long-range supersonic bombers capable of cruising at altitudes in excess of 60,000ft (18,288m).

The Americans began to investigate various options for future needs. Surface-to-air missiles remained in long-term development, but advanced manned aircraft could be built in the near term to plug the gap. Consequently, in 1951, two exotic prototype Republic XF-103 turbo-ramjet interceptors with a Mach 3.7 high-altitude capability were commissioned by the USAF.

This project was finally cancelled on 21 August 1957, but by then several supersonic fighters had reached an advanced state of development, including the Convair F-102A, the McDonnell F-101 Voodoo and the Lockheed F-104 Starfighter. Both the F-101 and the F-104 would enter RCAF service in the early 1960s, as the CF-101B and CF-104.

The British also considered a high performance interceptor using a mixture of rocket and jet propulsion, but this was finally abandoned and the supersonic English Electric Lightning fighter was introduced to counter the Soviet bomber threat to the UK.

At the start of the 1950s, Canada's government began to discuss the possibility of building an indigenous supersonic jet fighter. Ideally any new aircraft would have to be designed to operate from improvised sites

scattered across hostile northern regions of the country. This thinking had already played a pivotal role in the design of the RCAF's Avro Canada CF-100 Canuck fighter, which was able to take off from an improvised airstrip with a minimum length of about 4,000ft (1,219m).

Canada's aviation industry was developing rapidly and the government asked the Canadian Defence Research Board (CDRB) to consider every option for a locally designed and built interceptor. Subsequently, the CDRB approached Avro-Canada Ltd at Malton, Toronto (a subsidiary of A. V. Roe & Co Ltd of Manchester, England, and part of the Hawker-Siddeley Group). The CDRB requirement was stringent and they had to be certain that the new interceptor would remain effective well into the 1960s. Avro-Canada responded positively to the request and initial studies were soon under way, which received the name 'Project Y'. Although this was a Canadian programme, it was fully supported by the British parent company, while Whitehall and the Pentagon also secretly backed it.

This behind-the-scenes British and US interest in the project was hardly surprising, considering the close ties and shared defence interests of the three countries. CDRB Chairman Dr Omand Solandt had been Superintendent of the British Army's Operational Research Group during World War 2. He was in close contact with Sir Roy Dobson who headed Avro UK and was also a personal friend of Dr Vannevar Bush, the Pentagon's senior weapons technology advisor.

With everyone in agreement about the merits of proceeding with Project Y, a decision was made to begin formal start-up studies and the CDRB allocated $410,000 for the programme. Avro-Canada provided extra money for the project, assigned suitable facilities and then appointed the quietly spoken English engineer John Carver Meadows Frost as Project Y's director.

What was supposed to be a top-secret project soon became public knowledge, with Canadian newspapers reporting that Avro-Canada were working on a revolutionary flying saucer aircraft for the RCAF. In February 1953, *The Toronto Star* published a reasonably accurate description of the Project Y design and the story then found its way into an RAF Flying Review article entitled 'Man Made Flying Saucer'. From this point onwards, media interest in Canada's flying saucer project intensified

JOHN FROST & PROJECT Y

John Frost CEng, FCASI, MRAeS (1915-1979) grew up at Walton on Thames, England, and was educated at St Edward's School, Oxford. His interest in aviation was shared by the school's Latin teacher who was a qualified pilot. In 1930, he took Frost on his first flight in a Bristol biplane, although Frost would later recall that the experience proved rather unpleasant and he was unable to stop vomiting. It did not curb his enthusiasm for aviation, but Frost would have problems with airsickness throughout his life.

He began his aeronautical career as an apprentice with Airspeed Ltd, followed by a brief period with Miles Aircraft at Reading. In 1936, Frost moved to Westland Aircraft, working on the twin-engined Whirlwind fighter-bomber (which entered RAF service in 1940). This was followed by a job with Blackburn Aircraft at Brough in Yorkshire, mainly undertaking wind tunnel testing. He finally became a full-time designer with Slingsby Sailplanes where he produced the Hengist troop-carrying glider. The Hengist was a well-considered design, but it finally lost out to the Airspeed Horsa on the grounds of cost and simplicity of assembly.

Having met his future wife Joan (who worked in Slingsby's Tracing Office), Frost accepted a design job with de Havilland at Hatfield in 1942. He now worked directly for R. E. Bishop, participating in the development of the outstanding multi-role twin-engined wooden Mosquito, the Hornet, the DH.100

Above: **John Frost (left) and General A. G. Trudeau at a meeting in 1956.** *US Army*

Vampire jet fighter and the experimental DH.108 Swallow. The Swallow was a very advanced tailless swept-wing research aircraft, directly based on the Vampire. However, a great deal of wartime research undertaken by Dr Alexander Lippisch was utilised for the design. By spring 1946, the first of three prototypes (to fulfil Air Ministry Specification E.18/45) had been completed.

To prove a point about safety, Frost personally tested the Swallow's ejection seat and was launched up a 60ft (18.3m) high tower. This was something of an achievement for a man who suffered from airsickness. However, the Swallow project proved troublesome, partly due to the fact that it represented the cutting edge of aviation design that was being pushed to the limit. During a test flight on 27 September 1946, Geoffrey de Havilland was killed in the second prototype that suffered a structural failure above Egypt Bay in Kent. This would have a considerable impact on the programme, although the third DH.108 would become the first jet-powered aircraft to exceed Mach 1, on 6 September 1948.

Frost was now married to Joan and his first son Christopher was born in 1947. With the British economy in trouble, Frost decided to look for better work overseas and he accepted an offer from Avro-Canada to become their Design Manager for the CF-100

fighter programme. Frost quickly settled into the new job, but later admitted that he did not like the CF-100 very much, once referred to it as a 'Clumsy thing... all brute force'.

There were various design problems with the CF-100 and Frost set about correcting these shortcomings. He then began work on an improved swept wing version, although this never progressed very far beyond the concept stage as Avro-Canada were now considering the highly advanced CF-105 Arrow. In 1951, Frost was asked if he would be interested in heading Avro-Canada's top-secret Special Projects Group (SPG) who would develop a high performance VTOL fighter and he accepted. One of the conditions was that he and his family take Canadian citizenship, which was virtually standard procedure for all defence workers born overseas.

It remains unclear if Frost started out with a completely blank canvas when he became the head of Project Y, or was already working on ideas for supersonic VTOL interceptors, which seems plausible. He now recruited Thomas Desmond Earl and Claude John Williams for his team, who were both highly qualified aerodynamicists.

Frost had been interested in disc-shaped

aircraft for some years and studied the details of flying saucer incidents with considerable interest, keeping a large scrapbook of newspaper clippings and magazine articles in his desk. His initial reasons for choosing a disc-shaped aircraft for development remain slightly unclear, but it has often been suggested that Project Y was based on, or at least strongly influenced by, advanced German wartime research.

Immediately after the war, Frost is known to have studied captured aeronautical documentation and is thought to have been involved with the debriefing of several German technicians. It has not proved possible to track down any records of these discussions, but whatever he learned, it seems to have generated a major interest in VTOL aircraft. Throughout this period, Frost is understood to have produced a number of private concepts for helicopters powered by rotor tip jets.

The details of these designs remain unknown, but Frost's involvement with German wartime aeronautical research seems to have continued after he joined Avro-Canada. Some evidence of this came to light during searches conducted at Canada's National Archives. According to a document located by researcher Larry Koerner in the 1990s, an unusual meeting took place at a Canadian government facility in West Germany during 1953. On this occasion, Intelligence Officers from the RCAF, RAF and British secret service, accompanied by John Frost, met an unnamed German engineer who said he had worked on a secret flying disc project near Prague between 1944 and 1945. Apparently, the engineer claimed that a flying saucer shaped vehicle had not only been built but was test-flown. He said it was destroyed at the end of the war, along with all the documentation.

Also supporting the German connection is a document produced by W. E. Lexow, who headed the CIA's Applied Science Division and Scientific Intelligence Office. Dated 19 October 1955, it says, "Project Y is being directed by John Frost. Mr Frost is reported to have obtained his original idea for the flying machine from a group of Germans just after World War 2. The Soviets may also have obtained information from this German group." Was the initial Project Y aircraft directly based on captured German plans or details of an unusual engine design? If so, the details have never been revealed.

When Frost's small group of about eight designers and engineers started to consider options for Project Y, they looked at concepts for a single-seat flying disc with a diameter of approximately 40ft (12.1m). This aircraft would utilise an unusual circular-shaped jet engine that ducted its exhaust gases around three-quarters of the vehicle and used a system of segmented flaps for control. It would allow a full VTOL capability, hover and supersonic horizontal flight. However, initial engineering trials proved that the proposed system was impractical and would not function properly in VTOL mode. The team then considered the idea of adapting the design for a short rolling take-off and raised the front of the aircraft by 40° using two very long retractable struts, plus a tail wheel. This proposal was rejected by Frost, who decided to fully redesign the Project Y aircraft as a tail-sitter that would take off and land in an upright position.

The exhaust flaps were now discarded and engine gas would be ducted through a series of nozzles along each side of the aircraft and at the trailing edge. This configuration underwent considerable revision and directly affected the appearance of the aircraft, finally giving it an arrowhead or spade shape.

The link to earlier projects in wartime Germany remains unconfirmed, but one possibility is the propulsion system and it seems likely that this had been built and tested before design work started on the aircraft. There are strong hints that the unusual engine chosen for Project Y originated in Germany during the war and was developed by BMW for the elusive Schriever Project. Some sources have claimed that John Frost was directly responsible for designing the engine, but no

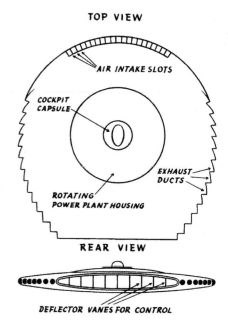

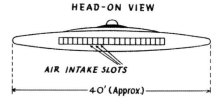

Above: **A drawing from the early 1950s that attempted to outline the main features of the Project Y design.** *Bill Rose*

Below: **An early detailed cutaway drawing of the Avro-Canada Project Y aircraft based on original design proposals.** *Avro-Canada*

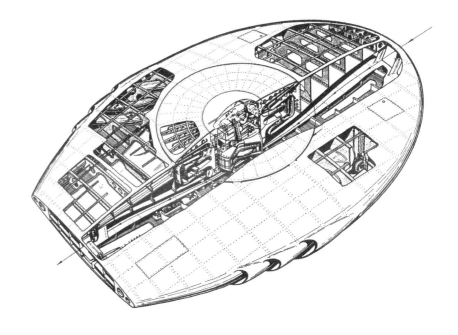

Flying Saucer Technology

Left: **Early artwork showing the possible appearance of the Avro-Canada flying saucer. This picture is thought to date from the early 1950s.** *Bill Rose*

Below left: **The Project Y aircraft, showing the location of the radial flow gas turbine engine and exhaust ducting.** *Bill Rose*

records have surfaced to support this possibility and it seems unlikely.

The unorthodox disc-shaped engine chosen for Project Y was called a Radial Flow Gas Turbine (RFGT). With conventional jet engines at an early stage of development and lacking power, the RFGT promised significantly performance gains. In rather crude terms, it can be described as a cylindrical jet engine stood on its tail and then flattened out into a disc. In this configuration, the turbine blades rotate 'edge-on' to the direction of flight. There is also a large compressor section and as the entire assembly revolves around the centre of the aircraft, it creates gyroscopic stability and would be expected to make the vehicle very steady as a weapons platform.

In the case of the Project Y design, air for the engine would be drawn in through two forward-facing intakes on the upper and lower fuselage. Compressed air would then be forced into a series of surrounding combustion chambers and through the turbine to a series of controllable outlets. Unfortunately, the true origins of the RFGT remain unclear, although there have been hints provided by documents sourced at the Public Record Office (PRO) in Kew, London (now called the National Archives).

If the engine was developed by BMW Bramo at Berlin-Spandau, then this technology almost certainly fell into British hands. Official documents indicate that post-war development of the RFGT was undertaken at the NGTE (National Gas Turbine Establishment) located at Pyestock, Hampshire, which was previously Frank Whittle's Power Jets Company.

It is also possible that development at Pyestock was conducted under the direction of Rolls-Royce's jet engine pioneer Dr Alan Arnold Griffith (1893-1963). Griffith was a brilliant scientist who developed advanced jet engine technology in parallel with Frank Whittle, favouring the more advanced axial flow compressor system to Frank Whittle's simpler centrifugal design. Frost was a personal friend of both Whittle and Griffith, with some unverified reports suggesting that Griffith provided technical assistance for Project Y in Canada. Engineering detail of the Project Y engine shows that the rotating assembly turned on roller bearings. This was

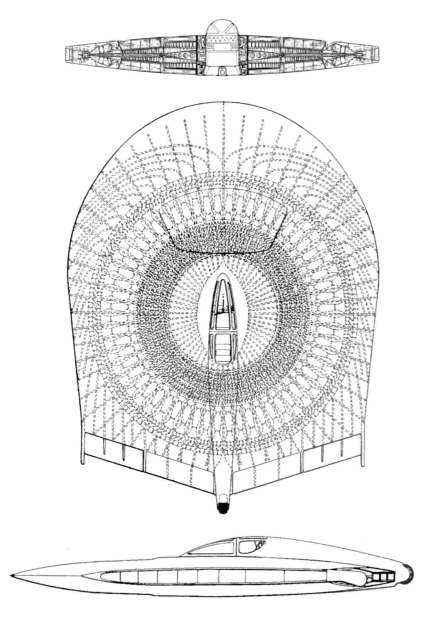

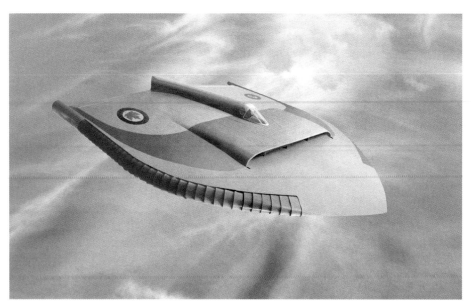

found to be unsatisfactory during development and was replaced by a more suitable air bearing arrangement.

Nothing is currently known about trials involving early prototype RFGT engines, specifications or performance, but work must have reached an advanced stage prior to approval of the Project Y interceptor. The proposed engine would have been quite substantial in size, with an estimated overall diameter of somewhere in the region of 25ft (7.62m). Furthermore the degree of integration with the airframe would make maintenance complex, costly and time-consuming. Additionally, there would be very little scope for future upgrades or improvements.

Specifications for the Project Y aircraft are often contradictory, but the production version was expected to attain a maximum speed of about Mach 2.25 in level flight, with an estimated service ceiling of 65,000ft (19,812m). A range of about 1,000 miles (1,609km) was anticipated with the aircraft in clean condition.

Dimensions are known to have changed during development and probably reflect alterations to the RFGT engine design and undercarriage layout. There is some disagreement, but the initial Project Y aircraft was expected to have a span of about 30ft (9.14m) and an overall length of approximately 40ft (12.19m). The span appears to have been reduced to around 25ft (7.62m) when the engineering mock-up was built.

No weights for the production aircraft are quoted in any available documentation. The centrally located cockpit carried the pilot in a conventional upright position and the issue of downward visibility was addressed by fitting two glass panels in the cockpit floor. Elevons and trimmers were located on the trailing edges and directed part of the exhaust flow.

Armament was expected to comprise two or four cannons and/or spin stabilised Folding Fin Aircraft Rockets (FFAR). At some later date, this might have been updated to two air-to-air missiles, carried on external rails.

Project Y was not designed to carry an intercept radar and it seems unlikely that adequate space could have been found to accommodate the bulky equipment of that period. Radar systems carried by aircraft generally required a second crew member and the RCAF appears to have decided that the CF-100 would continue to fulfil the night/all-weather role, while Project Y operated as a supersonic daylight interceptor directed from the ground.

Although Project Y was officially a Canadian programme, it received considerable unpublicised support from Avro at Manchester, England.

This included engine integration studies, wind tunnel testing with models, and test flights of at least one small experimental unmanned aircraft. This came to light when a previously classified USAF document from 1959 primarily discussing Avrocar mentioned that Avro UK had flown a small unmanned Project Y model aircraft in 1953. Unfortunately, no records of this exist as most of the Avro archives were destroyed during a major fire in the late 1950s.

Work progressed on the Project Y design in Canada, with the take-off and landing angle initially set at 75° upright, with a very long retractable undercarriage leg being used in combination with a small tail wheel. The arrangement was totally unsatisfactory and considerable effort was made to improve the system. This led to the aircraft being called the 'Manta' or 'Praying Mantis' by Frost's team, although the name 'Jump Gyro' can be found in some company documentation.

A full-sized engineering mock-up of Project Y had been completed at Malton and in early 1953 it received the official company name 'Ace'. The aircraft now stood on its tail in an upright 90° position, supported by a fully retractable strut with a wheel that was housed in the dorsal spine. There were two additional legs that extended from the underside, forming a tripod configuration.

It seemed that Frost's team was making good progress and Britain's Field Marshal Montgomery became the first visiting VIP to be shown the proposed interceptor. He was followed by British Minister Duncan Sandys, and it is evident that the UK was now taking

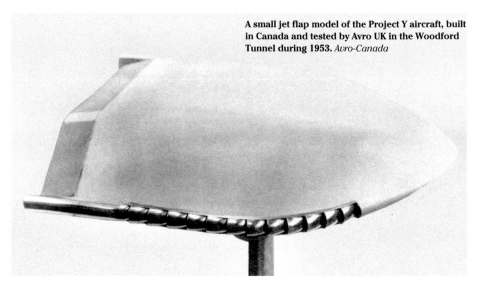

A small jet flap model of the Project Y aircraft, built in Canada and tested by Avro UK in the Woodford Tunnel during 1953. *Avro-Canada*

Flying Saucer Technology

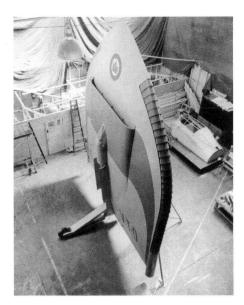

Above: **Sitting on its tail, the Project Y mock-up built by Avro-Canada's Special Projects Group in the early 1950s.** *Avro-Canada*

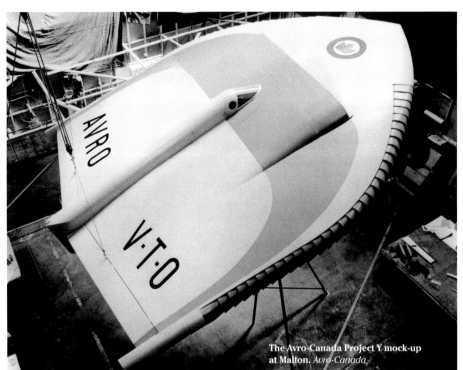

The Avro-Canada Project Y mock-up at Malton. *Avro-Canada*

quite an interest in the project. In March 1953, Dr Omand Solandt met Avro UK's Chairman Sir Roy Dobson and both men agreed that the Avro Ace showed considerable promise. However, Avro UK was working on a parallel VTOL interceptor with the company reference 724, which would take priority if an official requirement arose.

The 724 'tail-sitter' was designed to undertake a similar role to the Project Y aircraft, although it was slightly more conventional, using a high performance Rolls-Royce RB.106 axial flow turbojet engine that was still in the design phase. The anticipated static thrust of this advanced axial flow gas turbine was 15,000lb (66kN) and 21,800lb (97kN) using the afterburner. This was an integral feature of the RB.106 engine, using an automatically controlled convergent-divergent propelling nozzle. Although the RB.106 was cancelled in 1957, during defence spending cuts, it is understood that many engineering features of this engine were integrated into the Orenda Iroquois turbine developed for the Avro-Canada CF-105 Arrow.

The semi-dart-shaped 724 was more compact than the Project Y aircraft, with a wingspan of 16ft (4.88m) and an overall length of 32ft 6in (9.91m). The aircraft would be supported in an upright position by a single shock-absorbing strut housed in the single tail fin and two additional fully retractable struts

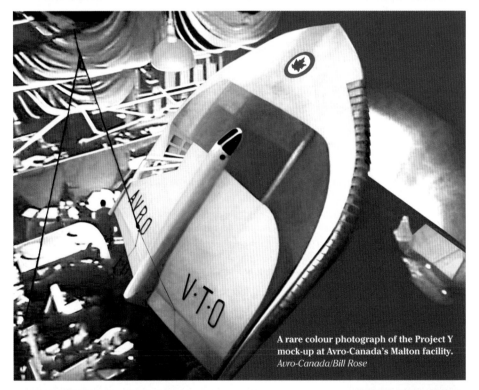

A rare colour photograph of the Project Y mock-up at Avro-Canada's Malton facility. *Avro-Canada/Bill Rose*

Right: **One idea explored by John Frost's team was to fit cockpits on each side of the Project Y aircraft. It is not clear how far this rather strange proposal progressed.** *Avro-Canada*

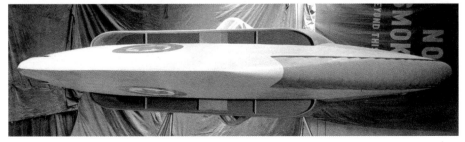

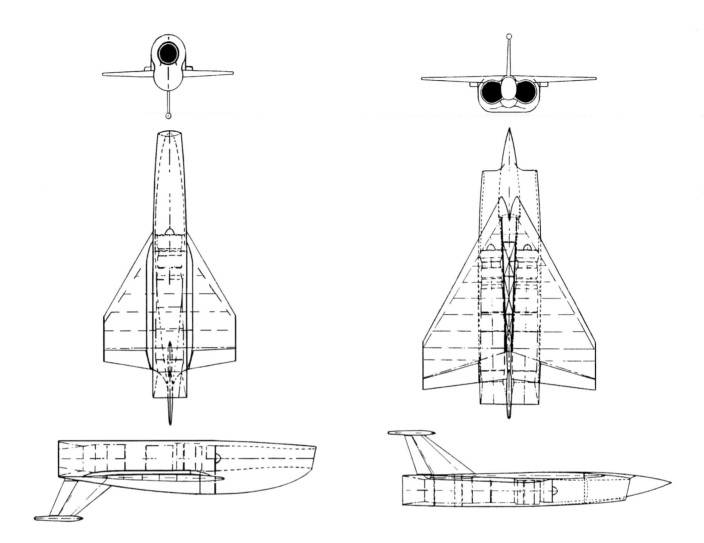

carried in fairings along the wing roots. The flush cockpit was located in a forward section of the lower fuselage, with the pilot in a prone position. Air for the jet engine would be provided by a nose intake.

Armament was not specified for this design, and the design would not have allowed a radar system to be carried. Details of performance are also unknown, although a maximum speed of at least Mach 2 was anticipated at altitude. As the project progressed, the design team became increasingly aware of the aircraft's limitations. As a VTOL interceptor, the 724 would be unable to attain an adequate thrust to weight (T/W) ratio with a single engine, and the lack of radar was an unacceptable handicap.

As a consequence, the aircraft was completely redesigned to accommodate two RB.106 engines, supplied with air from two inlets alongside the central pressurised cockpit which carried the pilot in a prone position. Now looking very much like a half-sized Avro-Canada Arrow, the aircraft was equipped with a 60° leading edge delta wing,

with a span of 24ft (7.32m). The overall length was 37ft (11.28m) and the quoted gross weight was set at 24,500lb (11,113kg). This included 8,000lb (3,628kg) of fuel carried in the wings and central tanks. Avro engineers believed this design would meet all the requirements for a VTOL interceptor, with a maximum speed of Mach 2.5 and a service ceiling of 60,000ft+ (18,288m+). The undercarriage was essentially the same as that used on the single-engined version of the 724, using a shock-absorbing strut carried in the tail fin's pod and two fully retractable struts carried in underwing fairings.

An unspecified airborne radar system was planned for this aircraft, although there are no details of how it might be carried, and proposed armament was two Blue Jay (later renamed Firestreak) air-to-air missiles mounted on wingtip launch rails. Soon after Dr Omand Solandt's meeting with Sir Roy Dobson it was decided to begin a series of wind tunnel tests with scale-sized 724 models. When news of this reached Canada, Frost was sent to meet with Sir William Farren of the Royal Aircraft

Above left: **An initial Avro (UK) proposal for the 724 VTOL interceptor. This design would have been powered by a single afterburning turbojet and it should be noted that the drawing is correctly oriented and not upside down!** *Bill Rose*

Above: **The second twin-engined proposal for the Avro (UK) 724 VTOL interceptor, showing some design similarities to the Avro-Canada Arrow.** *Bill Rose*

Establishment (RAE), in an attempt to sell him the idea of Avro-Canada's Ace becoming an RAF VTOL interceptor.

They finally agreed that the Avro 724 and the Ace would have similar performance and Farren accepted that the RFGT combined with a circular or delta-shaped aircraft provided several aerodynamic advantages, such as improved stall characteristics. Nothing conclusive emerged from this meeting in June 1953 and Farren remained non-committal. While efforts continued to generate British interest in the Ace, the RAF was shifting its attention towards a mixed-propulsion interceptor that would meet the latest OR.301 specification.

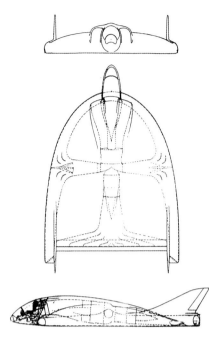

ducted into a peripheral plenum chamber that dispersed gases below the aircraft generating lift for take-off and landing. A letterbox-shaped exhaust vent at the rear was fitted with a flap to divert some of the exhaust downwards throughout VTOL operations to provide stability and additional control.

During transition to horizontal flight, internal doors would be used to progressively divert all the engine exhaust to the rear of the aircraft, and there would be small doors in the two outer exhaust ducts for additional yaw control.

Work continued to develop the Ace, but there were still serious problems with the undercarriage design. Attempts to resolve the undercarriage problem continued, but it remained clumsy and a potentially dangerous configuration and as a direct consequence the project was finally abandoned.

PROJECT Y2

Avro-Canada found itself in a difficult position with the Ace and there was no obvious solution. The CDRB had decided to scrap the entire project with immediate effect. They were already preoccupied with developing an aircraft to fulfil the new RCAF Specification 7-3, which would eventually lead to the superb but ultimately ill-fated Avro-Canada CF-105 Arrow supersonic interceptor.

The senior staff who had worked on Project Y must have been wondering if they would keep their jobs, but Frost had not entirely given up on the design. His team had made considerable progress with aerodynamic studies and engine integration. It was also clear that a completely circular 'flat riser' would be simpler to build and more likely to succeed.

Frost had been working on several alternative configurations for Project Y and wasted no time sharing these with senior management. However, there was the difficult issue of funding and there would be no more money coming from the CDRB. Fortunately, the Americans were still interested in the

Avro UK continued to develop the twin-engined 724 and also explored the idea of using a radial flow gas turbine engine in a triangular-shaped aircraft. Both studies were now scaled down, but apparently lasted for a further two or three years. Again this raises questions about how many RFGT engines had been built and tested by the early 1950s.

The 724 was not the only British high performance VTOL fighter concept in an early stage of development. A. A. Griffith at Rolls-Royce was working on a series of VTOL flat riser designs that included a two-seat, dart-shaped aircraft with a maximum speed of Mach 2.8, using a series of centrally located upright RB.108 turbojets for lift.

Hawker-Siddeley at Kingston also began to study options for a VTOL jet fighter, eventually producing a heel-shaped flat riser that may have been intended to compete with the later Avro-Canada WS-606A. This design is thought to have been produced by John W. Fozard, with work on the study ending in 1957 when drastic defence cuts were introduced.

No project reference was assigned to this Hawker-Siddeley concept, which would have been powered by a centrally located Bristol BE.23 turbojet engine. Exhaust would be

John Frost demonstrates the aerodynamic properties of his design in a laboratory.
Avro-Canada

design and the USAF, represented by Lt General Donald Putt, offered $200,000 to continue with Frost's work, with the promise of maintaining adequate financial support for the programme.

The British were clearly quite happy with this development. They had already reviewed the possibility of taking over the project, but decided that government R&D money could be put to better use elsewhere. Having received assurances that Avro-Canada would retain its rights and the RAE and Ministry of Supply would be given access to technical information, the US adoption of Project Y received official UK approval on the recommendation of Sir Roy Dobson and Sir William Farren.

Once the USAF took control of Project Y, a new office was opened at the Air Development Center within Wright-Patterson AFB, Ohio. In addition to making various USAF scientific facilities available to Frost's team, it

became possible to draw on special resources at the Massachusetts Institute of Technology (MIT) and NACA Ames in California. It also appears that security was stepped up and several US military personnel were assigned to the special Projects Group at Malton.

By mid-1954, the USAF was in full control of the Canadian project and several new designs were on the drawing boards. Meanwhile, the media remained interested in the Canadian flying saucer story and various reports continued to appear in the newspapers. The Canadian government now attempted to mislead the press with the admission that a disc-shaped fighter had been under development by Avro-Canada, but adding that the project had been scrapped due to ever-increasing costs. It was claimed that at least $100 million (several billion today) would have been needed to complete the project. While it was true that

the Canadians had abandoned the VTOL fighter, this was not the whole story. But the US media was reluctant to accept this explanation, with some correspondents suggesting that work was continuing in secret with Pentagon funding.

Finally, on 13 June 1955, the USAF admitted that they were funding an ongoing flying disc project at Avro-Canada and acknowledged a previously classified contract (Ref: AF33 (600) 30161) with the company. The USAF refused to discuss details of the programme, although USAF Secretary Donald Quarles followed this disclosure with the news that negotiations were already in hand to begin full-scale production of a revolutionary disc-shaped combat aircraft designed by John Frost. The exact reasons for releasing this information to the US and Canadian press remain unclear, but there may have been some underlying attempt to discredit Canada's Defence Production Minister, C. D. Howe. As a consequence, Howe was blamed for abandoning the flying disc project and allowing the Americans to take full control of it.

When previously secret USAF documents outlining the flying saucer project were declassified in 1995 under the US Freedom of Information Act (FOIA), they revealed two different saucer-shaped concepts. The programme had been assigned the name Project Y2 in 1954, but was also called Project 9961 Silver Bug within the USAF.

The two new designs produced by John Frost's group were derived from research carried out during Project Y, with the first completely circular compact proposal being intended to act as a demonstrator. Performance of this VTOL design was subsonic and the overall diameter was 21ft 6in (6.55m).

This research aircraft was too small to carry the RFGT engine and in its place were eight radially dispersed Armstrong-Siddeley Viper axial flow turbojet engines. Each would draw air from inlets around the centre of the aircraft and their exhausts would be ducted towards the edge of the disc. During take-off/landing and hover, the exhaust gas would be directed downwards to create a powerful lifting effect and this would be redirected in level flight.

The cockpit was located at the upper

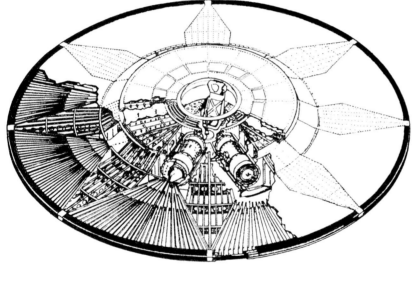

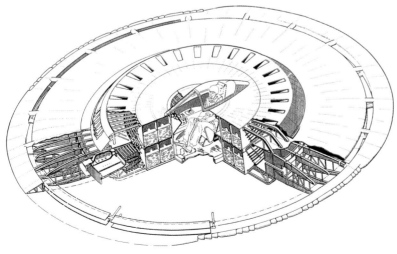

Above left: **First of a series of advanced proposals that followed the ill-fated Project Y design, this small eight-engined experimental research aircraft was intended to act as a proof-of-concept demonstrator for the Silver Bug project.** *USAF*

Left: **A sectional drawing of the advanced Project Y2 Silver Bug aircraft powered by a second-generation radial flow gas turbine.** *USAF*

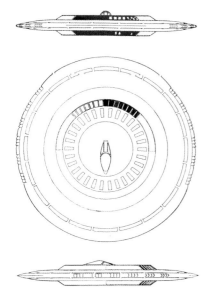

centre of the aircraft and covered by a one-piece dome-shaped canopy. Although very cramped, an ejector seat was fitted. That aside, there would have been serious issues with visibility, and a glass panel was set in the floor below the pilot's legs to improve matters. The second more advanced design was much closer to a production aircraft capable of becoming a high performance interceptor or high-altitude reconnaissance vehicle. This design was sometimes referred to by the secret USAF codename Project Ladybird and was powered by an improved RFGT engine.

The RFGT for this aircraft was under development by Orenda Engines (Avro-Canada's jet engine division) and formed an integral part of the aircraft. It drew air through upper central inlets for VTOL operation and front-facing upper and lower inlets for forward flight. One significant improvement over the earlier engine was the use of an air bearing to support the compressor/turbine wheel. The combustion chambers were coupled to an afterburning system, which ducted gas to a complex arrangement of exhaust vents. In comparison to a conventional gas turbine engine, there was more to go wrong and maintenance would be costly and very labour intensive. Nevertheless, the estimated performance of this design was nothing short of sensational, with an anticipated maximum speed of Mach 3.5 and a service ceiling in excess of 80,000ft (24,384m). The aircraft was expected to reach 70,000ft (21,336m) in 4.2 minutes from a hovering position. Today, there are no combat aircraft with the ability to come anywhere near this level of performance.

The aircraft's overall diameter was 29ft (8.84m) and its gross weight was calculated at 29,000lb (13,154kg). A constituent of this was 950 UK gallons (4,318 litres) of fuel, providing a maximum range of about 620 miles (997km). Despite the periods of sustained high-speed flight being relatively limited, there would still be issues of thermal build-up that could not be avoided. In 1955, the best available metals for airframes and skins were less than ideal for this purpose and it was hoped to partly overcome the problem by dumping excess heat into the central fuel tanks.

Flight control would be achieved by selective direction of the engine's exhaust and using the Coanda Effect to bend the gas flow. Pitch and roll would be regulated by the annular nozzles, and the trailing edge exhausts would control yaw. There were several modifications considered to the overall system, including the addition of small louvred exhaust vents to improve directional control, although it appears they were not included in the final design.

A simpler proposal for flight control was the use of an annular ring that surrounded the entire aircraft and channelled the engine's exhaust gas to selected outlets. This had been devised for the research aircraft, but was found to perform poorly during wind tunnel tests of models. At supersonic speed there was excessive drag and in level flight the lower engine intake started to perform badly, due to exhaust gases flowing across the bottom of the aircraft. The schematics for the research

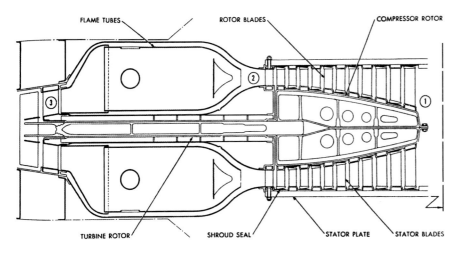

Fig. 10 Typical Cross-Section Through Engine

Top: **General appearance of the RFGT-powered Project Y2 aircraft.** *USAF*

Middle: **Project Y2 engine exhaust angles at low speed.** *USAF*

Left: **Cross-section of the Project Y2 Radial Flow Gas Turbine (RFGT) engine.** *USAF*

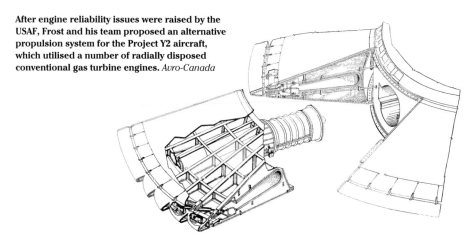

After engine reliability issues were raised by the USAF, Frost and his team proposed an alternative propulsion system for the Project Y2 aircraft, which utilised a number of radially disposed conventional gas turbine engines. *Avro-Canada*

aircraft and the larger RFGT-powered design show no provision for an undercarriage, but it seems almost inevitable that retractable landing legs would have been incorporated to avoid damage to the underside of the aircraft and allow easy ground handling.

In 1956, the USAF began to express serious concerns about the reliability and combat survivability of the RFGT engine. From the very beginning of Project Y, it had been obvious that in the event of an engine failure, the aircraft would become impossible to fly or land. These resurrected issues led Frost and his senior designers to reconfigure the Ladybird aircraft to accept an alternative propulsion system.

This consisted of 6 or 8 radially dispersed jet engines, positioned around the central cabin in a pattern similar to the small research aircraft. Frost felt certain that they could provide adequate power without a serious weight penalty. The airframe would be designed so each engine was accommodated in an individual section with accompanying fuel tanks and each of these units could be removed fairly easily for maintenance purposes. Outwardly, the new variant looked almost identical to the RFGT-powered aircraft, and because of general improvements to gas turbine engines the performance was expected to be very similar.

Something of an enduring puzzle is how Frost expected to arm this aircraft, although it is known that there were proposals to carry two or four cannons. Exactly how these weapons would have been installed is unclear, as space within the aircraft was extremely limited. One possibility considered by Frost was to fit a heavily armoured leading edge to the aircraft, allowing it to ram enemy aircraft at high speed. The idea had already been considered by designers in wartime Germany and is said to have been suggested by Jack Northrop during development of the P-79 rocket-powered flying wing fighter. The use of a flying disc for this purpose was also briefly explored in the 1949 movie serial *Bruce Gentry – Daredevil of the Skies* and Frost felt that the idea was an effective and serious possibility.

He believed that it would be possible to smash through an enemy bomber at speeds in excess of 2,000mph (3,218kph) with virtually no prospect of damaging the disc and very little risk to the pilot. Whether this would have been the case remains unknown, but finding a fighter pilot willing to test the theory was probably another issue! Some official documents also indicate that there were suggestions that this aircraft might be used as a light bomber. In this role, stores could only have been carried beneath the aircraft and this is a further indication that a fully retractable undercarriage was planned.

The USAF intended to use the aircraft for the defence of North America, and both the RCAF and RAF remained open to the idea of future acquisition if the project went well. In addition, the US Navy also expressed considerable interest in Project Y2, recognising its potential as a VTOL fighter capable of operating from full-sized aircraft carriers, small ships and modified submarines, equipped with fully enclosed hangars.

Frost studied UFO reports with considerable interest and suspected the Soviets were already operating flying disc aircraft from ships and adapted submarines to make clandestine flights over US territory. This seems to reinforce the belief that a Project Y2 variant was considered in exactly this role as a high-altitude and totally deniable spyplane.

Other variants studied by Frost's team included a much larger RFGT-powered aircraft with a diameter of 100ft (30.5m) and a central thickness of 20ft (6.1m). It would be configurable as a long-range bomber, a reconnaissance vehicle or a transport aircraft. Whether or not this design would be powered by a large RFGT is unknown, but a maximum speed of Mach 3, with a ceiling of 90,000ft (27432m) was envisaged, with a cruise altitude of 65,000ft (19,812m) and a range of 15,000 miles (24,140km).

Two further (known) developments of the RFGT-powered flying disc were considered under the Project Y2 Silver Bug programme. Both were remote-controlled guided missiles and the first was an unmanned version of the Ladybird aircraft that would almost certainly carry a nuclear warhead in place of the pilot. The second was a miniature flying disc with a diameter of 6ft (1.8m). This was intended for use on the battlefield, either as a missile or as a reconnaissance vehicle.

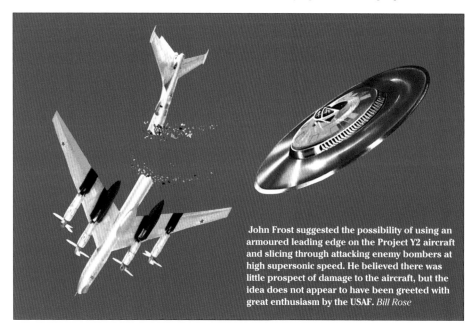

John Frost suggested the possibility of using an armoured leading edge on the Project Y2 aircraft and slicing through attacking enemy bombers at high supersonic speed. He believed there was little prospect of damage to the aircraft, but the idea does not appear to have been greeted with great enthusiasm by the USAF. *Bill Rose*

PROJECT MX 1794

As work began on Project Y2 Silver Bug in Canada during 1954, the USAF began to develop an alternative design at the Air Research & Development Command, Wright-Patterson AFB. This was known as Project MX 1794 and, rather confusingly, the 1794 designation had already been used on some documents for Frost's RFGT aircraft. The USAF prefix MX is normally associated with experimental missiles, but appears to have been applied to a number of different classified programmes and was simply an abbreviation of 'Military Xperimental'.

Project MX 1794 was expected to deliver a much higher level of performance than the Silver Bug aircraft, due to an advanced and rather ingenious turbo-ramjet propulsive system. With ongoing concerns about the reliability of the RFGT engine, MX 1794 would become the principal aircraft under development by Frost's Special Projects Group at Malton. Although the USAF had developed the MX 1794 design, the propulsion system can be traced back to 1953, when Frost and Williams first considered alternatives to the RFGT.

The initial concept for this study envisaged a flying disc with a diameter of approximately 35ft (10.7m). It was powered by a series of radially dispersed axial flow gas turbines that were used to turn a large compressor stage, which revolved around the aircraft's upper central section, generating gyroscopic stability. The compressor forced air through a series of ducts to flame holders around the circumference of the aircraft and the exhaust gas was finally channelled through shuttered ports that could be adjusted to provide flight control.

After transition from hover to level flight, the pilot would throttle back the turbojets, allowing the ramjet function to completely take over. A conventional and relatively simple ramjet can only work at high speed, but this design was very different and could function at all speeds and while the aircraft hovered. Two variations of this proposal allowed different engine arrangements and alternative cockpit layouts. However, the fourth study combined all the best features of these designs and the Air Research & Development Command chose it for parallel development with Project Y2.

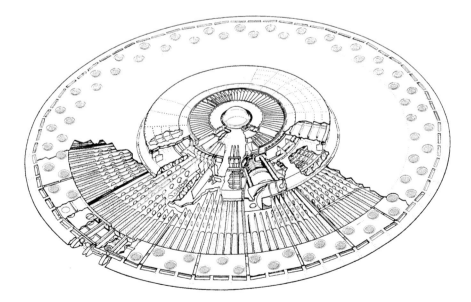

Above left: **One proposal was to modify a number of US Navy submarines to carry Project Y2 aircraft.** *USAF*

Below left: **An early design that would lead to the MX 1794 turbo-ramjet-powered flying disc.** *Avro-Canada*

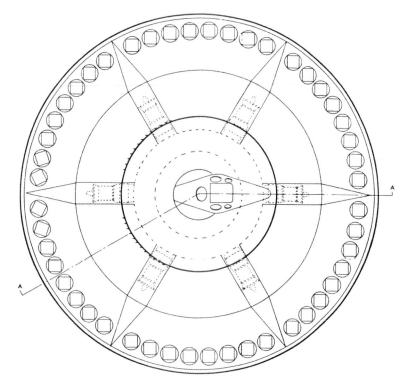

UPPER IMPELLER AND TURBINE
UPPER AND LOWER INTAKES
FLAME HOLDERS
OUTER WING
INNER WING DIFFUSER SECTION
LOWER IMPELLER AND TURBINE
PILOT'S COCKPIT
INTEGRAL FUEL TANKS
ENGINE INTAKE
FLIGHT CONTROL SHUTTERS
6 A. S. M. VIPER 8 ENGINES
TURBINE EXHAUST
ENGINE TAILPIPE

SECTION A-A

1648-1794-1

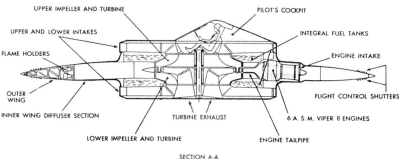

FLIGHT CONTROL SHUTTERS
PILOT'S CANOPY
ENGINE ACCESS PANEL
UPPER INTAKE
INTEGRAL FUEL TANKS
EXHAUST DIFFUSER
TRIM METERING VALVE
FLAME HOLDERS
INTERNAL AIRFLOW DUCTS
ASM-VIPER 8 ENGINE
ENGINE AIR INTAKE
IMPELLER
TURBINE
LOWER INTAKE
INNER WING
OUTER WING

6-VIPER DUCTED FAN RAMJET DESIGN

Left: **Originally classified, this drawing shows the principal features of the Avro-Canada MX 1794 six-engined, turbo-ramjet-powered flying disc interceptor.** *USAF*

Below left: **Key features of the Avro-Canada MX 1794 interceptor.** *USAF*

In early 1955, Project 1794 (the MX designation had now been discarded) replaced the Project Y2 Silver Bug programme and Quarles allocated a further $750,000 in research funding. By September, Frost's engineers completed a 1/6th-size scale model with a diameter of about 5ft (1.5m) which was shipped to Wright-Patterson AFB for wind tunnel testing. At the same time, work began at Malton on assembly of an engine test rig.

Six Armstrong-Siddeley Viper turbojets would be used, with their exhausts facing inwards to drive a pair of 8ft (2.44m) diameter contra-rotating centrifugal impellers that powered the ramjet system. The Vipers provided about 9,000lb (40kN) of static thrust, but the overall ramjet system was expected to produce 17,800lb to 32,750lb (79kN to 145kN) thrust.

When the first prototype was tested, it was planned to restrict the exhaust temperatures to 926°C, providing a Mach 1.6 capability. This would have been progressively uprated to 1,426°C, allowing a speed in excess of Mach 3.7 at extreme altitude. It was anticipated that the production aircraft would be able to attain Mach 3.5-4.0 in level flight and operate at altitudes as high as 105,000ft (32,000m). The rate of climb was also impressive and expected to be no more than 3 minutes to 75,000ft (22,860m) from hover. The aircraft would easily attain Mach 2.25 at 90,000ft (27,432m) and cruise at this altitude for 37 minutes. With a proposed gross weight of 27,322lb (12,393kg), Frost estimated the aircraft's range at 1,000 miles (1,609km).

Dimensions of the 1794 aircraft were now set at 35ft 3.5in (10.76m) in diameter and 7ft 8.5in (2.35m) from the lower surface to canopy top. The cockpit canopy was of a more robust design than the previous RFGT aircraft and would have been better suited to sustained high-speed flight, although visibility for the pilot was reduced due to use of extra metal sections.

As with any discplane having a centrally located cockpit, there were some serious downward visibility restrictions. In an attempt to address this issue, it was suggested that a periscope system could be used, although it would have been a complicated design. It was also decided that the aircraft needed to be equipped with three fully retractable landing

Flying Saucer Technology

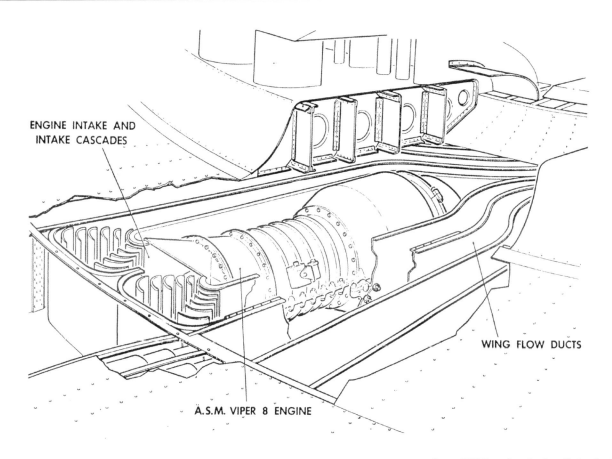

ENGINE INTAKE AND
INTAKE CASCADES

WING FLOW DUCTS

A.S.M. VIPER 8 ENGINE

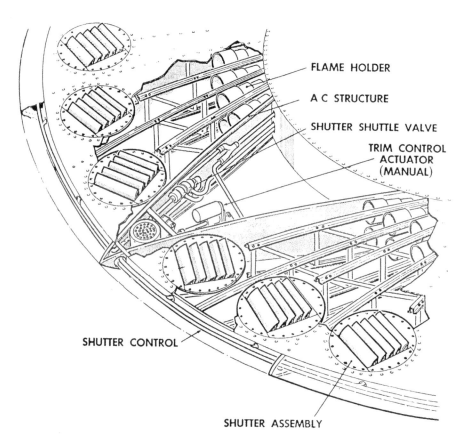

FLAME HOLDER

A C STRUCTURE

SHUTTER SHUTTLE VALVE

TRIM CONTROL
ACTUATOR
(MANUAL)

SHUTTER CONTROL

SHUTTER ASSEMBLY

Above: **ASM Viper 8 engine installation for the Avro-Canada MX 1794 flying disc.** *USAF*

Left: **The relatively complicated exhaust shutter control system proposed for the Avro-Canada MX 1794 aircraft.** *USAF*

legs which would be fitted with wheels and tyres to aid ground handling. The aircraft would be 2ft (0.6m) above the ground, which might have proved inadequate for the carriage of external stores. On the other hand, it is possible that this figure (found in official documentation) refers to clearance beneath the lower central section of the aircraft.

As the project continued to evolve, the scientists at Wright-Patterson AFB expressed concern about the propulsion system and the fact that the aircraft was totally dependent on engine power for flight, stability and control. While the failure of one turbojet might have been tolerable in an emergency, the overall reliability of the turbo-ramjet system remained a matter of some concern.

Edward H. Schwartz, who worked for the Aircraft Laboratory, said the following in a classified evaluation of the 1794 aircraft: "Should a complete power failure or exhaustion of fuel occur, it would be impossible to make a forced landing. Whether sufficient controllability would exist even with ram-air at high forward speeds, to enable the pilot to establish favourable

Left: **The 6-Viper test rig at Avro-Canada, Malton, Toronto.** *USAF*

within the cockpit.

When trials of the engine rig commenced in early 1957, this became apparent and was followed by several small fires and serious reliability problems, highlighted when one of the six Vipers ran wild, creating a very dangerous situation.

The question of how to effectively arm the 1794 aircraft underwent considerable revision, but the idea of ramming enemy bombers at high speed does not appear to have been considered feasible. The initial plan was to install a 20mm Gatling gun, and then it was decided to supplement this with two externally carried air-to-air missiles (AAMs). These would have been AIM-7 Sparrows, AIR-2A Genie rockets with nuclear warheads or long-range AIM-26A nuclear-tipped Falcon AAMs. This might have being the simplest arrangement, but the carriage of external weapons or supplementary fuel tanks was considered unsatisfactory by the USAF, as the aircraft's performance would have been degraded.

There was also the question of how to equip the 1794 with a radar system, and classified studies were undertaken by the Hughes Aircraft Company and Ramo-Wooldridge Corp in 1956. They determined that the radar/fire control system for 1794 would weigh a minimum of 2,055lb (932kg) and require 41cu ft (1.16m3) of space. How this equipment was to be installed is unknown, although it probably involved the use of a forward nacelle. In its final configuration, the interceptor was expected to weigh approximately 32,000lb (14,514kg).

An officially undisclosed variant of the 1794 aircraft was considered for high-altitude daylight photo-reconnaissance missions. It would be fitted with a camera system similar to the unit carried by the Lockheed U-2 spyplane. This 1794 variant would possess a number of obvious advantages. It would be able to cruise at extreme altitude and was expected to be very stealthy. Dick Cantrell of Lockheed's Skunk Works once remarked that a flying saucer represented the ultimate shape for low observability and this must have been fairly obvious to designers in the late 1950s. Furthermore, it would be possible to discredit any chance sightings of the aircraft as bogus UFO reports. Other versions of the 1794 are believed to have been considered, which included an interdictor or light bomber (with ordnance carried

ejection conditions is questionable." Nevertheless, this does not seem to have had any impact on the project and by December 1956, the engine test rig, which represented roughly three-quarters of the aircraft, had been completed in a special facility at Malton.

When trials of the Orenda turbo-ramjet system started, concerns were expressed about thermal stress and the use of special combustors that had been fabricated in England. The suitability of metals for various components was now in question, especially in areas such as the exhaust system controllable outlet vents and it was

acknowledged that new high-temperature alloys would need to be found. It was also clear that high supersonic performance would heat the skin and parts of the airframe to substantial temperatures that might prove unacceptable. In response to this, it was decided to manufacture certain parts of the airframe from temperature-resistant Nimonic (Nickel-Chromium) alloys and fabricate the aircraft's entire skin from 300-series stainless steel of 0.43in (10.92mm) thickness. US scientists were also becoming worried about noise, with the Vipers expected to produce 140dB, which would be totally unacceptable

Principal features of the 6-Viper test rig. *USAF*

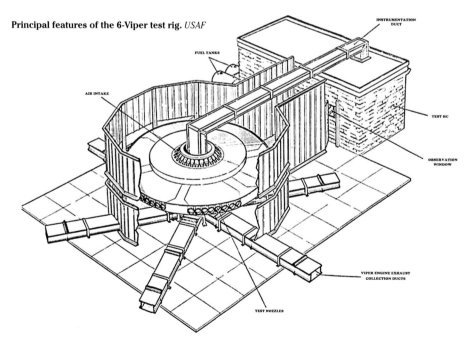

INSTRUMENTATION DUCT

FUEL TANKS

AIR INTAKE

TEST HC

OBSERVATION WINDOW

VIPER ENGINE EXHAUST COLLECTION DUCTS

TEST NOZZLES

Right: **Avro-Canada engineers inspect and adjust the central component of the 6-Viper test rig during the final stage of assembly.** *USAF*

Below: **The central section of the 6-Viper test rig is lowered into position using a crane.** *USAF*

externally) and an unmanned aircraft, perhaps intended for use as a missile.

There are still many unanswered questions about Project 1794 and the Project Y RFGT-powered aircraft. Exactly how far did any of this work progress and could it have been secretly passed to a US contractor? During a project review at Wright-Patterson AFB on 4 November 1955, Avro-Canada's Vice President Fred T. Smye made it clear to officials that his company was ready to begin assembly of the first 1794 prototype.

Smye told the review group that development was progressing so well he would be asking the Avro Board to provide $4.5 million to ensure construction of the first aircraft. USAF staff then determined that it would cost approximately $50 million to fully develop the 1794 aircraft by 1961, which was considered a perfectly acceptable figure for any 1950s high-performance aircraft.

In late December 1955, the USAF was advised that around three-quarters of the aircraft had been built and in July 1956, USAF Major J. W. Frost (no connection to John Frost) from the Wright-Patterson Air Development Center visited Avro at Malton to discuss the construction of a second prototype. Wright Air Development Center had now completed their wind tunnel testing of the 1/6th scale model and ground effect studies were taking place.

Right: **Artwork showing a USAF Avro-Canada military aircraft based on the MX 1794 operating from an improvised site.** *Avro-Canada*

Above: **The MX 1794 full-plane wind tunnel model used during trials at MIT.** *Avro-Canada*

Left: **The 1/6th scale half-plane model used for wind tunnel testing.** *Avro-Canada*

The preliminary report raised a number of concerns and this was followed on 1 August 1956 by the results of separate wind tunnel tests on models at MIT. Each set of trials had determined that the lift/drag (L/D) ratio of a circular planform was unsatisfactory at low subsonic speed and could not match the most sophisticated swept wing design at supersonic speed.

The controlling USAF office accepted these findings but remained convinced that the outstanding thrust/weight ratio provided by the turbo-ramjet propulsion system totally compensated for any aerodynamic shortcomings. By September 1956, wind tunnel testing was being undertaken at NACA's Ames facility in California and Project 1794 resumed. During a presentation to the US Navy at the Pentagon on 11 October 1956, Admiral Russell was advised that the first Project 1794 prototype was scheduled for completion by January 1957. What was really going on is far from clear as documents appear to suggest that Project 1794 was abandoned soon after this.

WS-606A

The Special Projects Group (SPG) were also working on an alternative to Project 1794, which was a multi-role supersonic VTOL combat aircraft that would be offered to the RCAF. This was known as Private Venture 704 (PV.704), but rather confusingly was also briefly given the title Project 1794, which then became Performance Research System Nr 453L. The USAF and CIA began to seek a successor to the high-altitude Lockheed U-2 spyplane in 1955 and PV.704 immediately caught their attention. They realised that this design might provide exactly the kind of performance that was required for the next USAF spyplane.

The first version of the PV.704 was very similar to the 1794 aircraft, although it was equipped with a dome-shaped, mostly metal cockpit canopy with viewing ports. Many design reviews followed and PV.704 finally evolved into something quite different. This concept would use a new central fuselage section containing conventional jet engines combined with an enclosed rotor system that provided lift. The aircraft was fitted with a conventional forward cockpit section for the pilot, with internal storage space for fuel or extra equipment.

The USAF was pleased with the way that PV.704 was progressing and substantial black budget funding was made available to continue with development. This also meant that the project was renamed as Weapons System 606A (WS-606A).

Subsequently, two versions of the design emerged under WS-606A. The first was a single-seater called Configuration A, powered by six side-by-side Armstrong-Siddeley Viper jet engines, fed by large over-wing air intakes. The large rotor at the centre of the aircraft would duct air to the periphery of the wing for lift during take-off and landing. However, as the project progressed, it became evident that the aircraft's endurance could be substantially improved if a rolling take-off was used. Described as Ground Effect Take-off (GETO), the calculations indicated that using this method would increase the mission range by around 35 per cent.

Specifications for the Configuration A aircraft were revised many times, but the following figures appear to be fairly accurate. The wingspan was 29ft (8.84m) and wing area is quoted as 1,083sq ft (100.6m²). The

overall length was 37ft (11.28m) and height was 7ft 8in (2.33m). Quoted weights are also difficult to verify, but the Configuration A aircraft powered by six Vipers was expected to have a gross weight of 65,000lb (29,483.5kg).

A variant of the Configuration A design using four larger turbojets was also considered, but this appears to have been the only notable design difference. The initial performance estimates for WS-606A suggested a maximum speed of Mach 4 at an altitude of 95,000ft, (28,953m), but this appears to have been the disc-shaped design, which was similar to the Project 1794 aircraft. Later figures for Configuration A indicated a maximum speed of Mach 2.5 with a ceiling of 65,000ft (19,812m). The combat radius would have been 800 miles (1,287km), with a ferry range of 3,300 miles (5,310km).

An alternative design for the WS-606A aircraft was given the designation Configuration B and this became the preferred choice. Although approximately the same size with a similar wing, the tailless fuselage was completely different and accommodated two unspecified afterburning jet engines. It was estimated that the total thrust produced would be 55,000lb (244kN). Air for the engines was drawn through a nose inlet using a conical centre body, and this version was designed to carry a pilot and navigator in tandem cockpits.

The aircraft would be supported on the ground by three sets of double wheels with an absolute minimal amount of clearance beneath the wing. This restriction might have been considered necessary for GETO, but would have created many problems that ultimately proved insoluble. With a similar gross weight to the Configuration A design, the performance estimates were now revised downwards to show a more realistic maximum speed of Mach 2 and a combat radius of about 500 miles (804km) using VTOL and approximately 700 miles (1,126km) with GETO.

As an interceptor, Configuration B would

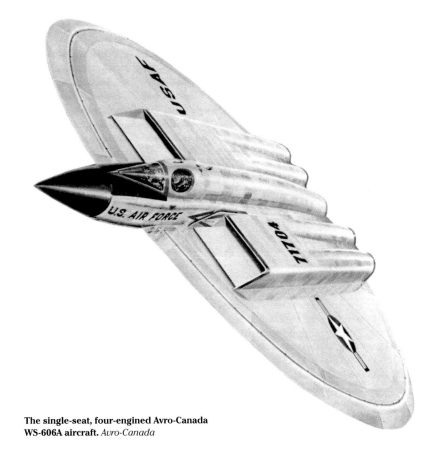

The single-seat, four-engined Avro-Canada WS-606A aircraft. *Avro-Canada*

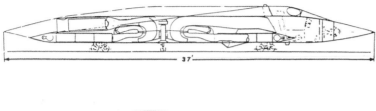

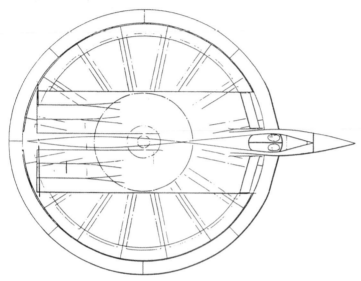

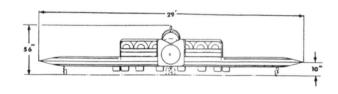

The six-engined Avro-Canada WS-606A proposal.
USAF

have been equipped with two air-to-air missiles (AAM) and a suitable radar system. However, USAF documents quote a modest payload of 1,000lb (453kg) for this aircraft, and the AAMs and radar would have exceeded this figure by a considerable margin. Configuration B was apparently also considered for the role of a light-bomber or attack aircraft, although the payload capability seems too restrictive for a conventional mission or the carriage of a single 1950s-era nuclear weapon.

Perhaps the real issue is exactly how Frost's team or the USAF envisaged the carriage of any ordnance. There appears to be insufficient space for a forward bay and the minimal ground clearance would have prevented any attachment below the aircraft. It is conceivable that plans existed to carry two AAMs on launch rails above the wing, which was a method used for the Anglo-French Jaguar strike aircraft. On the other hand, it is more likely that WS-606A was never seriously considered for anything more than daylight photo-reconnaissance.

Security for Project WS-606A was organised by the USAF and highly compartmentalised. One consequence of this was a complex paper trail, intended to mislead anyone gaining unauthorised access to documentation. This had made research into the work carried out by Avro-Canada's SPG extremely difficult and very time-consuming. One example was use of the WS-606A designation for the Avrocar project, which became active in 1958.

While this complication may have been considered worthwhile, it clearly caused

considerable difficulty for USAF personnel who appear to have often been confused by the use of WS-606A in connection with Project 1794 and the Avrocar. Several declassified documents show requests for clarification on project designations, which appear very muddled. Eventually the problem was recognised and WS-606A was referred to as the Supersonic Application, while using the name Avrocar for the following project.

In January 1958, John Frost advised the USAF Project's Office that a full-sized prototype of WS-606A had been assembled, minus the outer wings. This is confusing, as the wings would seem essential to the design. No pictures or documentation of this aircraft have surfaced and exactly what was being described remains a puzzle.

The following month, the question of further funding was raised by Avro-Canada and the USAF were told that one completed WS-606A supersonic prototype would cost three times the price of two smaller Avrocar subsonic utility vehicles. The following month, a 1/12th scale model of the supersonic design was transported to MIT for wind tunnel tests. Exactly what was taking place at this point remains unclear, but there was growing disagreement between the USAF and Avro-Canada about a number of technical issues.

This finally came to a head on 22 March 1958 when Joseph Ellis, who controlled the Design Branch of the USAF's Aircraft Laboratory, recommended that all work on the WS-606A supersonic aircraft project should be immediately stopped. The project hung in the balance for many months and this situation was finally overtaken by problems with Avro-Canada's CF-105 Arrow which was unexpectedly cancelled. Many employees were laid off on 20 February 1959, including Frost's team and it looked as if Avro-Canada was finished. But three days later, all members of the SPG were reinstated to continue work on the WS-606A supersonic aircraft and Avrocar.

Their first objective was to deal with the main matters raised by USAF scientists: correcting wing flutter and sonic fatigue issues. The USAF continued to support WS-606A and estimated that the cost of this project would reach $50 million by 1965, when it was anticipated that the aircraft would be ready to enter service. Research continued in Canada and the US until summer 1960, when the Pentagon formally cancelled the project, having decided that it no longer had any need for this aircraft.

With the benefit of hindsight, it would seem that the WS-606A supersonic aircraft had little to offer anyone apart from a limited

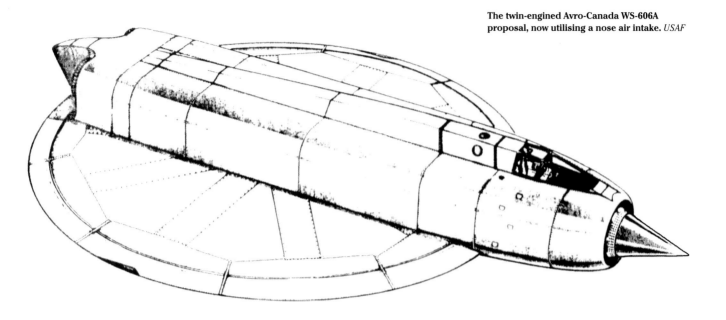

The twin-engined Avro-Canada WS-606A proposal, now utilising a nose air intake. *USAF*

VTOL capability. As a piece of machinery it was extremely complex, making the aircraft unattractive for operations at improvised sites. Performance was adequate rather than outstanding and there were problems with the payload capability. By contrast, the superb Avro-Canada CF-105 Arrow long-range interceptor would have been far superior to WS-606A in most respects, apart from lacking a VTOL capability.

The Arrow should have entered RCAF service and perhaps been adopted by the RAF. There was great potential for further development: proposals included a Mach 3 variant, capable of attaining an altitude of 100,000ft (30,000m), a long-range, high-altitude photo-reconnaissance aircraft and an interdictor. But unlike the Arrow, which was cancelled for political reasons, it must have been obvious at an early stage that the WS-606A was technically flawed and had little potential for future development. The USAF office responsible for controlling this project also funded development of several other exotic and ill-fated programmes, which included the hydrogen-fuelled Lockheed CL-400 Suntan, which was developed to replace the U-2.

It soon became clear that the Suntan was a mistake and it was scrapped. Then, in 1959, Lockheed's Mach 3+ A-12 Blackbird was chosen to succeed the U-2. However, rather ironically, the Blackbirds are now in museums, whereas (at the time of writing) some later versions of the U-2 remain in service. Full histories of the WS-606A and CL-400 Suntan remain undisclosed and it is generally believed that these costly and unsuccessful black projects are sources of lingering embarrassment.

AVROCAR

In 1957, John Frost turned his attention to the design of a smaller, multi-purpose subsonic version of the 1794 concept, making several approaches to the USAF and US Army with various proposals.

By the following year he had produced a more refined design that finally secured a development contract from the US Army. Coincidentally, there was media speculation that Avro-Canada had started work on a new flying saucer known as VZ-9, which was surprisingly accurate and suggests a carefully controlled official leak. The new aircraft was known as the VZ-9AV Avrocar and official disclosure took place in Washington during the April 1959 Congress Space Committee Hearings. Few details of the Avrocar were released, although it was noted that this was not a space vehicle of any kind.

After completion in May 1959, the first Avrocar (AV-7055) underwent some ground testing and was sent to NASA Ames for wind tunnel trials in November 1959. At the same time, a poor quality overhead photograph of a disc-shaped vehicle standing on the tarmac at Malton was made available to the press, with Flight Magazine describing it as a 'sneak shot'.

Confusingly, the Avrocar started out as a USAF project designated WS-606A, but the

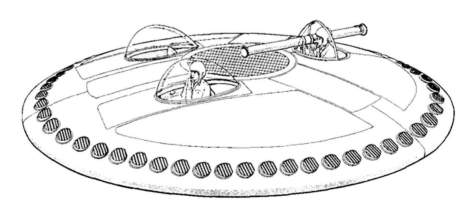

This is one of many concept drawings for early versions of the Avrocar known as Avrojeep. *USAF*

Left: **An early overhead photograph of Avrocar at Malton.** *Avro-Canada*

VZ-9AV reference was applied once it became a US Army programme, despite the USAF remaining in control of the project.

The Avrocar utilised some technologies developed during the earlier projects, but differed considerably from any of the high performance proposals. Avrocar was designed to function as a compact, two-man air jeep or utility craft with full VTOL and hover performance. Its maximum speed was expected to be 300mph (482kph) at an altitude of 10,000ft (3,048m) and it would have a range of 1,000 miles (1,609km), carrying a 2,000lb (907kg) payload in the rear bay.

The Avrocar was 18ft (5.49m) in diameter, 3ft (0.91m) high and weighed 5,650lb (2,562kg). Propulsion was provided by three symmetrically disposed Continental J-69 turbojets, each rated at 927lb (4.12kN) static thrust. The exhaust from these gas turbines would drive a centrally located fan with an overall diameter of 5ft (1.52m) turning at 2,780rpm. This was expected to generate gyroscopic stability and dampen any pitch and roll motions. The exhaust was then ducted to nozzles around the periphery of the vehicle, to allow lift and forward flight. Soon after testing began, many serious shortcomings became apparent. The air jeep was little more than an incredibly noisy hovercraft that did not function properly,

John Frost and T. Desmond Earl must have been aware that there were problems before the first tests, and documents declassified in 1990 show that the Wright Research and Development Center already had major concerns about Avrocar. Clearly the biggest issue was the propulsion system, with unacceptable turbine inlet temperatures preventing the J-69s from reaching optimal efficiency. In turn, this prevented the central fan from turning at maximum speed and meeting performance requirements. It is also said that heat within the cockpit was so high that it damaged some of the instruments and pilots had to wear protective suits.

There was no possibility of Avrocar meeting any of the early performance estimates. It was not only incapable of leaving the ground but initially almost impossible to control. To sum up in one word, the Avrocar was a disaster.

While the USAF and US Army prepared to abandon the programme, Frost and his team hurriedly undertook design revisions that addressed most of the problems. This led to the Mk 2 Avrocar, which was fitted with a side-by-side, forward-positioned cockpit. The Mk 2 had the same 18ft (5.49m) diameter but was powered by two General Electric J-85 gas turbines driving a 6ft (1.83m) diameter fan, with some central exhaust ducting to improve performance. Four short hydraulically dampened landing legs would support the vehicle at rest. As the Mk 2 evolved, it was decided to attach a tail fin, and a further development was fitted with short swept wings that had stabilising fins at the tips.

It is worth mentioning that many variants of the Avrocar were studied by Frost's team, including several commercial designs, called Avromobile, and remote control vehicles for military use, known as Avrodrone and

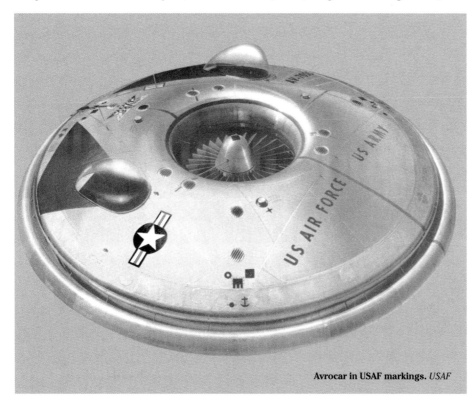

Avrocar in USAF markings. *USAF*

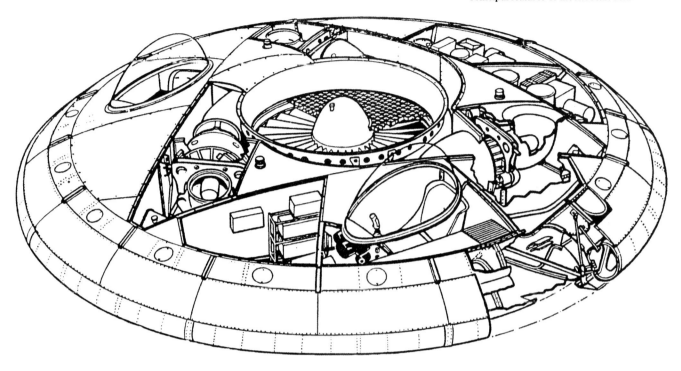

Avropelican, which both looked rather like the later Sikorsky Cypher. Some of Frost's engineers would later claim that the Avrocar could have worked if more money had been made available. But the Pentagon thought otherwise, and in June 1961, all USAF involvement with Avro-Canada ceased and Frost's special projects were at an end.

In fact, the fate of Avro-Canada had already been decided when an ill-considered decision was taken to cancel the CF-105 Arrow. Subsequently all the best staff were recruited by US contractors or NASA and in early 1962, the company was shut down and the facility was sold to Douglas Aircraft of Canada. Today, only a couple of original buildings remain and the site has been replaced by an industrial area attached to Toronto Airport. It appears that three Avrocars were assembled and enough components were fabricated to produce a fourth. The third Avrocar was broken up for scrap along with all the other parts.

The Avrocar tested and demonstrated at Malton can now be seen at the US Army Transportation Museum, Fort Eustis, Virginia. The other Avrocar shipped to NASA for testing

A brief test flight at Malton of the second VZ-9AV Avrocar in 1958. *Avro-Canada*

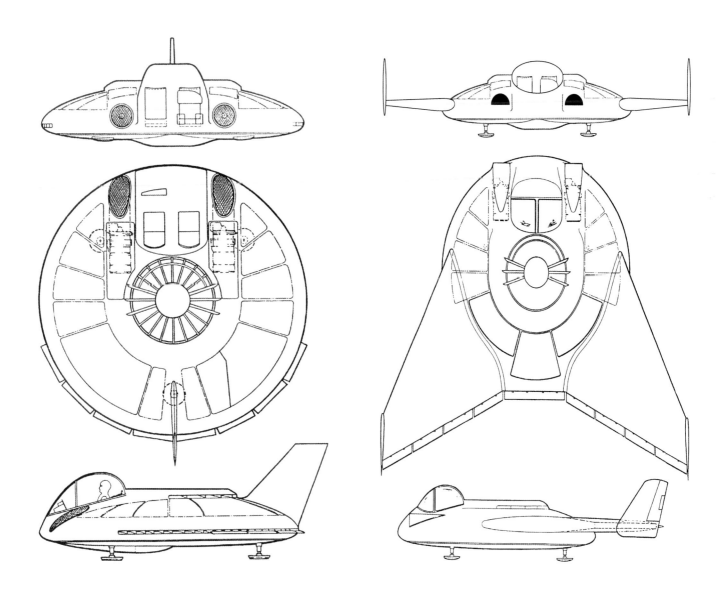

Above: **The greatly improved Avrocar Mk 2, now powered by two engines and utilising a tail fin.** *USAF/Bill Rose*

Above right: **This is the final and perhaps most advanced Avrocar design. Based on the Mk 2 proposal, it had the same body size at 18ft (5.4m), with a wingspan of 29ft 3in (8.9m), a length of 24ft 6in (7.46m) and a height of 6ft 6in (1.98m). This compact aircraft was to be powered by two General Electric J-85 turbojets and it would utilise rudders and a jet flap system for horizontal flight control. No details of anticipated performance are known.** *Bill Rose*

was passed to the National Air & Space Museum and currently resides in the Paul E. Garber Storage and Restoration Facility at Silver Hills, Maryland.

Although Frost discovered the ground cushion effect during Avrocar's development, the credit went to British designer Sir Christopher Cockerell who is widely

acknowledged to have invented the hovercraft. It is said that Frost was less than happy about this and became deeply disillusioned with the aircraft industry. Once Avro-Canada folded, Frost accepted a job with the Airworthiness Section of the Civil Aviation Administration in New Zealand. He briefly returned to Canada in 1964, becoming involved with a boat business in Vancouver, but this did not work out and he returned to New Zealand, being appointed Technical Services Engineer for Air New Zealand.

During his time with the company, Frost was responsible for several engineering innovations and improvements. These included a tail docking system that enhanced maintenance and engine changes for DC-10 aircraft. After his retirement in 1978, Frost continued to design aircraft and was working on a small man-powered glider when he died on 9 October 1979.

WERE THE SOVIETS DEVELOPING FLYING SAUCERS?

John Frost took considerable interest in UFO sightings, often studying the reports in great detail and trying to determine the probably cause. While it seems that he was not sold on the extraterrestrial hypothesis, Frost is known to have voiced the possibility that flying saucers might be long-range Soviet spyplanes, perhaps launched from the decks of merchant ships. Their primary mission would be to look for weaknesses in American air defences. At this time, the Cold War was moving progressively towards a military confrontation and the idea that some flying saucers might be Soviet spyplanes was unofficially popular with a number of senior Pentagon and CIA staff.

From the late 1940s, there were genuine concerns that the Russians were in possession of advanced German flying disc technology that might give them an edge in any major conflict. It would also provide an explanation for some of the more reliable UFO reports. Might there have been a Soviet equivalent to the Avro-Canada programme that had already led to operational aircraft? Although the CIA's intelligence gathering capability steadily improved during the 1950s, much of what took place inside the Soviet Union went unobserved during this era and as a consequence there was endless speculation. The Soviets had fully utilised advanced German technology to improve their military capabilities and were working on a range of projects from jet fighters to ballistic missiles derived from the V-2.

Some evidence that the Soviets were testing new weapons seemed to come from Scandinavian sightings of bright fireballs called Ghost Rockets, which began in 1946. The Swedish government initiated an official investigation, which involved British Intelligence, but it was finally concluded that these events were nothing more than natural phenomena. In 1948, something similar started to happen in the United States, with Green Fireballs being reported in the night skies above New Mexico. The USAF took a serious interest in these sightings and appointed Dr Lincoln La Paz, who was the Director of the University of New Mexico's Institute of Meteoritics to investigate. He decided that because the Green Fireballs followed flat trajectories and moved at relatively low velocities, they were almost certainly artificial in origin.

Having submitted his initial report, it was immediately classified as top secret, although the media had been alerted to the issue, making the information difficult to suppress. Pentagon officials were now considering the possibility that Green Fireballs were evidence of advanced long-range Soviet vehicles crossing sensitive regions of America at high altitude. Almost immediately, Dr Joseph Kaplan from the University of California at Los Angeles (UCLA) was commissioned to undertake a second study.

He completed his report on 3 November 1949, determining that although the objects were decidedly unusual, they were meteors of natural origin and had been observed because of excellent 'seeing' conditions in New Mexico. Kaplan also mentioned heightened public awareness of unusual activity in the night sky due to UFO reports in the press. The Ghost Rockets and Green Fireballs may well have been natural objects reported by over-eager observers, but it seems likely that nobody was sure what they were dealing with and it was decided that

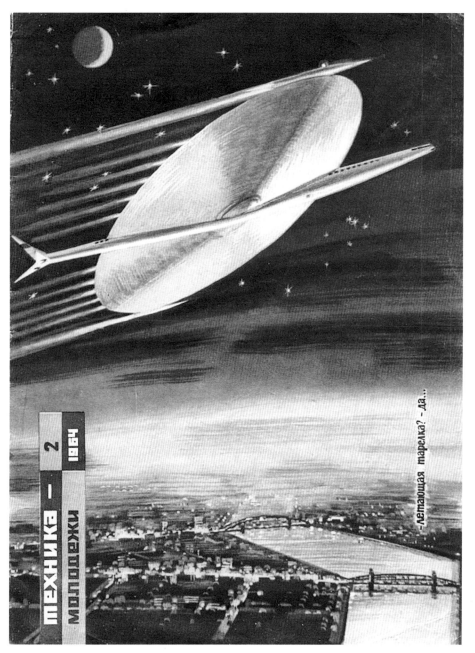

Above: **Soviet magazine artwork from 1964, depicting a futuristic airliner with a circular wing. This had no connection to any ongoing project and was purely fictional.** *Bill Rose*

public interest needed to be suppressed, pending further secret investigation.

After Dr Kaplan's report the issue faded away, but three years later Dr La Paz opened up the debate once more by claiming that Green Fireballs seen around the world were a new type of missile, conceivably built by the Russians. On 28 June 1952, the West German newspaper Saarbrücker Zeitung reported claims that a wrecked Soviet flying saucer had been found on the remote Arctic island of Spitsbergen (Svalbard).

Apparently a flight of six Norwegian military jets had experienced severe radio interference while flying over an area of the island and pinpointed the cause as a crashed flying disc that was partly buried in the snow.

Almost immediately, several teams of specialists were dispatched from Narvik to secure the craft and undertake an initial inspection. One of the first people on the scene was a Norwegian rocket specialist called Dr Norsel who determined that radio direction finding equipment powered by a small plutonium cell had been responsible for transmitting the jamming signal.

The preliminary survey showed that this vehicle had been unmanned. No crew members were present and the craft had been flown by remote control. Evidently something had gone wrong with the control system, causing it to crash. The disc had a diameter of about 160ft (48.8m), was made from unknown metals and used 46 jets around the vehicle's circumference for propulsion. There was a central control section with instruments marked with Russian letters, and investigators determined that the machine was capable of reaching an altitude of 100 miles (160km) and had a near global range. It was later determined that the disc was probably based on wartime German technology.

This story appeared in *Berliner Volksblatt* on 9 July 1952 and *Der Flieger* in August 1952, and then overseas publications took the story, which was being handled by the AFP news agency. As the report evolved, it would be claimed that the crashed saucer had been travelling to the United States on a spying mission and similar vehicles were responsible for many worldwide UFO sightings. Nevertheless, the entire story was total nonsense. There were many factual errors and the originator(s) of this detailed fabrication were never traced.

While the report was clearly a hoax, it does begin to look as if this was intentional disinformation rather than mischief. The Russians were incapable of building anything even remotely like the alleged Spitsbergen saucer, but the timing was perfect for renewing Pentagon fears about advanced Soviet weapons. Furthermore, there can be little doubt that this will have helped to secure funding for continuation of advanced US projects like the Avro-Canada flying saucer. On the other hand, it remains possible that there were Russian attempts to develop disc-shaped aircraft based on German research and perhaps a plan to duplicate the Avro-Canada project.

Both Schriever and Epp insisted that the flying saucers being sighted around the world were Russian in origin, with Epp claiming to have worked on the construction of an early prototype in East Germany. He described the vehicle as being based on one of his proposals, but less complex, using fixed position RD-10 (copies of Junkers Jumo 004B) jet engines to replace the ramjet rotor assembly. Apparently the Russians also modified his lift system, using six ducted fans as opposed to eight.

This new disc-shaped aircraft was fitted with two vertical stabilising fins, a centrally located cockpit and a fully retractable four-strut undercarriage. According to Epp, the Soviets planned to use this aircraft for specialised reconnaissance operations and the prototype was completed at Pirna (near Dresden) during 1950. He was unable to provide any further details, but believed the aircraft had entered

Above: **The Sukhanov Discoplan 1 on landing approach. This small one-man glider was demonstrated at Tushino in 1958 and its initial purpose was to pave the way for a larger disc shaped, jet-powered utility craft. This never happened and Discoplan 1 is currently on display at the Russian Federation Air Force Museum at Monino.** *Pete Schneider/Bill Rose*

limited service. Epp seems to have spent much of his later life claiming that others had taken his ideas and developed them without his knowledge or compliance. While researching this particular story, I was unable to find any real supporting evidence. However, on 5 October 1957, the official State magazine *Sovetskaya Rossiya* briefly referred to an unusual aircraft designed by Professor S. Zonshtein. It was described as a 'circular object with four ducted fans'. On its own, this story means nothing, but I am reluctant to completely dismiss Epp's claims, which may seem improbable but remain interesting.

Much the same can be said about a supposed secret Russian programme based on work undertaken by the Prague group. It is possible that this project existed, but the difficulties in locating official documentation in Russia can be considerable and no supporting evidence has come to light. The known developers of Russian post-war discplanes were a small team headed by Mikhail Sukhanov. His design bureau does not appear to have been given an official OKB designation, but it was responsible for a series of gliders, possibly ground effect vehicles and several exotic concepts between the early 1950s and mid-1960s.

The first Sukhanov design to fly was a very compact single-seat glider called Discoplan 1. It was built at Novosibirsk and demonstrated at Tushino in 1958. Discoplan 1 consisted of an

almost circular wing with a diameter of 11ft 6in (3.5m) and an area of 107.6sq ft (10.2m²). Take-off weight is quoted at 551lb (250kg) and maximum speed was limited to 80mph (130kph). The aircraft was fitted with an upper tail unit with a rudder and there were two large control surfaces on the wing. The pilot sat in a pod-shaped cockpit beneath the wing and the undercarriage comprised two fixed forward wheels and a single skid beneath the wing's trailing edge.

An examination of the aircraft shows a poor standard of construction and a serious lack of attention to detail. Discoplan 1 appears to have been produced to test ideas for a small disc-shaped, jet-powered utility aircraft and is currently on display at the Russian Federation Air Force Museum at Monino. Sukhanov followed this design with Discoplan 2, which was completed in 1960 at the Lavochkin plant. The wingspan is thought to be 16ft 5in (5m), but the exact overall length is uncertain.

There is a forward fuselage section with a well-positioned cockpit affording good visibility, and the undercarriage consisted of a

semi-recessed wheel directly below the cockpit and two skids on the wing's trailing edge. Discoplan 2 was fitted with an all-moving semicircular tail fin, plus two large trailing edge

Top left: **Forward view of the Sukhanov Discoplan 2 glider. Completed in 1960, the exact purpose of this aircraft remains unknown.** *Alexi Malinovsky*

Top right:**The Sukhanov Discoplan 2 glider prepares to land. It has been suggested that this design was used to develop a powered circular-winged aircraft, but no documentation has surfaced to support this.** *Alexi Malinovsky*

Below: **Rear view of the Sukhanov Discoplan 2 glider.** *Alexi Malinovsky*

control surfaces. The purpose of Discoplan 2 is unknown, although it may have been built to test the properties of a more advanced powered design. There may have been a Discoplan 3, but I have been unable to locate any details for this.

The mysterious Sukhanov Design Bureau was also responsible for producing a compact disc-shaped, propeller driven seaplane and this development group was clearly much more than just a small unit constructing experimental gliders.

ALTERNATIVE RFGT RESEARCH

At least two other designers are known to have taken a serious interest in developing RFGT engines for aircraft propulsion. One of these individuals was American inventor Bedford A. Smith, who filed a US patent application on 18 April 1955 for a Turbine Powered Convertible Aircraft.

In appearance his aircraft was semi-disc-shaped, with a rear body section. The dome-shaped cockpit would be centrally located on the upper side of the aircraft's centre and subject to some visibility restrictions.

The RFGT engine design was very much a part of the aircraft and it used a multi-stage centrifugal compressor and an estimated 32

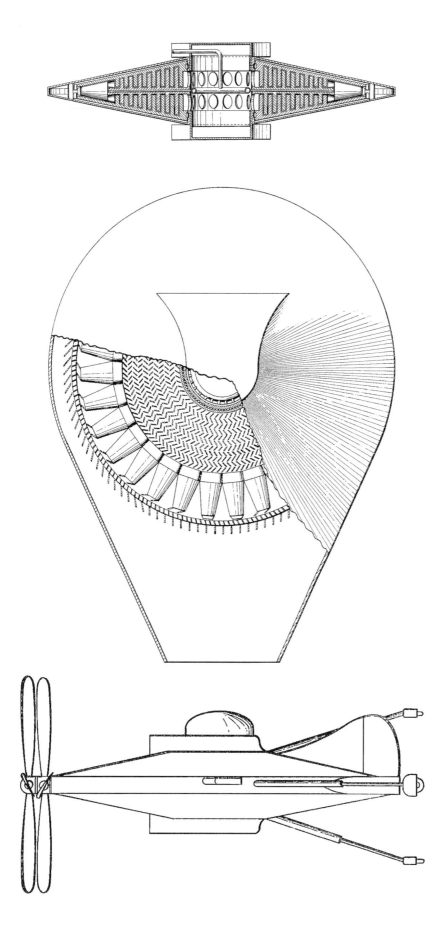

Top and middle: **An unusual design for RFGT-powered semi-disc-shaped aircraft produced by Bedford A. Smith in the mid-1950s.** *US Patent Office*

Bottom: **A variant of the RFGT-powered aircraft design created by Bedford A. Smith, which utilised a shaft-coupled forward set of contra-rotating propellers and operates as a tail-sitter.** *US Patent Office*

combustion chambers, with the turbine revolving on roller bearings. Fuel would be supplied via a feed in the central rotor hub and air would be directed to the engine via upper and lower forward-facing inlets. Smith also considered the possibility of shaft coupling the turbine to a set of contra-rotating tractor propellers. In this configuration, the aircraft would operate as a VTOL tail-sitter, supported on four fully retractable struts with wheels.

Flight control would be achieved using side nozzles that would act like ailerons. Smith also intended to direct the flow of exhaust gas leaving the rear of the aircraft. The design was surprisingly accomplished and seems to have been intended for a military role. At the present time, relatively little is known about Bedford Smith's background and although his aircraft was never built, the design is often cited in later patents and has proved influential.

HEUVEL'S SAUCER

When the American designer Norman L. Heuvel started design work on a flying saucer in the 1970s, he examined all the previous engine and aerodynamic concepts and decided to pick out the best features and make some obvious improvements.

In many ways, the aircraft he developed could best be described as a modern version of Frost's Silver Bug, with a sleek shape that looks as if it belongs in a SF movie.

Powered by an advanced RFGT, Heuvel improved the basic design by starting out with a twin rotor system that would completely eliminate torque and gyroscopic precessional forces. Air for the engine would be drawn in around the cockpit and through the compressor section. However, both rotors would utilise the same annular combustion section. Another innovation was the use of electromagnetic rotor braking to provide yaw control during flight. Exhaust gas from the engine would be ducted downwards during VTOL operation and rearward during horizontal flight, with directional control assisted by six spoiler flaps on the upper and lower surfaces of the aircraft.

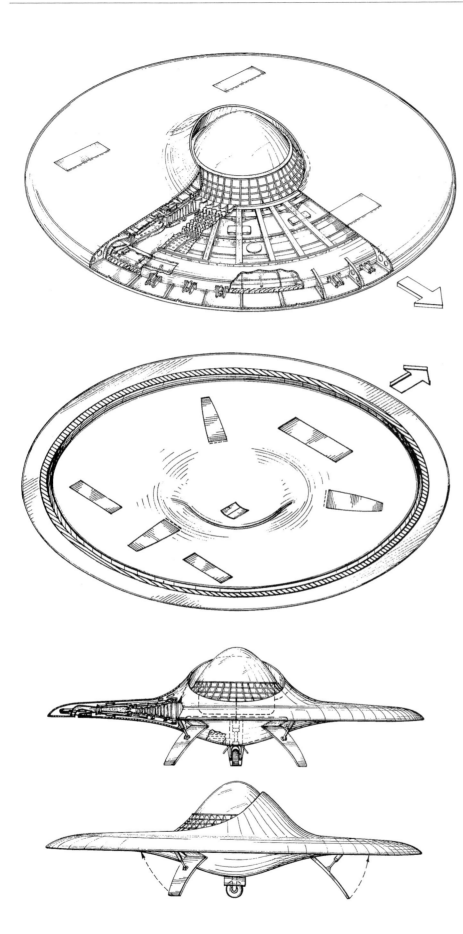

Top: **Developed in the early 1970s by Norman Heuvel, this design for an advanced flying saucer is powered by a radial flow gas turbine engine and builds on earlier work undertaken by engineers like John Frost.** *US Patent Office*

Middle: **The underside of Heuvel's flying saucer, showing control surfaces and the positions of undercarriage/ground support features.** *US Patent Office*

Bottom: **Forward and side view of the Heuvel flying saucer, showing a cross-section of the radial flow gas turbine and the landing gear extended.** *US Patent Office*

Heuvel claimed that the aircraft would be much quieter than a conventional jet aircraft due to the exhaust gas being released as a thin sheet. This elegant design would be supported on the ground by a single retractable wheel below the centre of the aircraft and three contoured landing struts. Heuvel applied for a patent in July 1976 and reapplied just over a year later. This was published in 1980 (4193568) and he continued to refine the design, but failed to find any financial backing for his aircraft.

LOCKHEED SPECIAL PROJECTS

Although Lockheed (now Lockheed-Martin) has always tried to distance itself from discussion of concepts such as flying saucers, this is something that its special projects division, otherwise known as the Skunk Works, studied in the 1950s.

Lockheed may have felt obliged to develop plans for a flying saucer aircraft in response to work being undertaken by Avro-Canada. Heading up this programme was Nathan C. Price, a leading propulsion expert, who had already studied the designs produced by Weygers and Loedding in some detail.

By 1952, his team had developed plans for an impressive high performance ramjet-powered saucer with a VTOL capability. Consequently, several related US patents were filed on 23 January 1953 and these remaining classified for around the next ten years. (The most significant US patent has the reference: 3103324.) There are US Patent Office regulations to handle designs and inventions that are considered sensitive to national security and these were undoubtedly applied to this very detailed material.

It is interesting to note that one of the first patents to be released shows Nathan C. Price as a resident in Mexico City, while the second says Hollywood, California. This is typical document alteration to create a false paper trail and it indicates the classified nature of

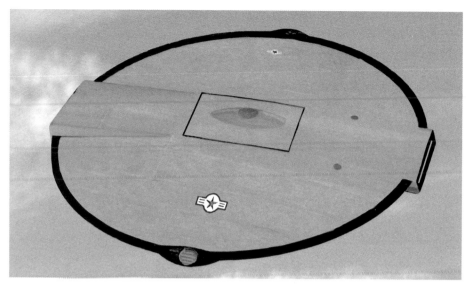

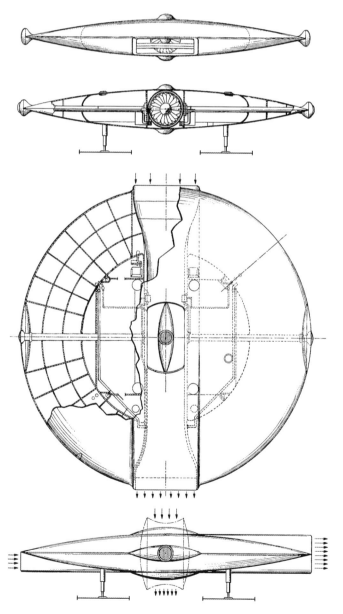

the project. Furthermore the documents show only a civil application for the design, with accommodation for a small number of passengers. In the author's opinion, this was simply an attempt to conceal the true nature of the design, which was purely military.

Lockheed's flying saucer was undoubtedly conceived as a high performance light bomber or reconnaissance aircraft with a level of performance far in excess of the North American XB-70 Valkyrie developed in the following decade. It was probably expected to operate from improvised bases and penetrate ever-improving Soviet air defences in the pre-emptive strike role.

In the available documentation for this project, Price remarks that the bi-convex planform was a rigid and strong structure, allowing an even landing force distribution, and it would provide maximum volumetric payload efficiency and cost-effective construction. His flying saucer was compact and capable of full VTOL, plus high supersonic speed at extreme altitude. Propulsion took the form of a large centrally positioned duct, with a carefully configured variable area inlet feeding a powerful turbo-ramjet propulsion system.

The central engine unit, described as an 'Island', would swivel downwards during take-off and landing to provide lift. Initially, this unit was intended to contain a single turbojet derived from one of Price's early wartime designs, but it was replaced by a cluster of more advanced engines.

Having made the transition to level flight, the aircraft would climb under turbojet power, assisted by afterburning, and at supersonic speed the propulsive system would become a ramjet. The exhaust gases would be channelled through a series of vanes at the rear of the aircraft that would provide pitch control. Air from the engine island would also be ducted to small shaft-driven (load) turbines to provide directional control and stability.

A number of different fuels were considered for use with this aircraft: aside from kerosene, these included propane, butane and liquid hydrogen. Fuel would be circulated beneath critical areas of the aircraft's skin, acting as a refrigerant. It was calculated that stainless steel that covered

Top: **Representation of a Lockheed high-altitude VTOL ramjet-powered flying disc developed for the USAF to undertake nuclear strike or reconnaissance missions.** *Bill Rose*

Left: **General appearance of the Lockheed high performance VTOL flying disc designed by Nathan Price's team during the early 1950s.** *Bill Rose*

Flying Saucer Technology

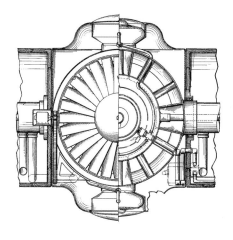

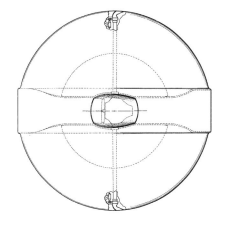

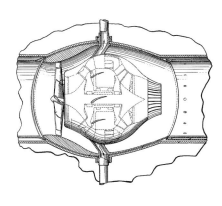

the aircraft would glow a dull cherry red during sustained flight at Mach 4, having reached a temperature of 615°C. Price believed this could be reduced to an acceptable level by dumping excess heat into the fuel and also by using an external black finish to improve thermal emissivity.

The Lockheed lenticular-shaped aircraft was almost completely circular, with a diameter of 50ft (15.2m), making it very compact for an aircraft offering such a high level of performance. The gross weight was quoted at 55,000lb (24,950kg) and while the payload weight is not stated, it is possible to make some deductions based on available information. In a military role, it could have carried at least two free-fall nuclear weapons (of that era) or several high resolution cameras for a photo-reconnaissance mission. Available documentation also talks of a pilotless variant, which may imply that the aircraft was considered for use as a missile.

As a special forces transport it might have accommodated 18 fully equipped soldiers, although I should stress that these suggestions for applications are purely speculative. In the role of a civil transport aircraft, the Lockheed flying disc would have provision for 24 rearward-facing passengers and their luggage in two separate cabins. The aircraft would be flown by a crew of two, located in separate cockpits on either side of the engine inlet. It is not clear if there was allowance for a navigator, flight engineer or cabin staff (in the civil proposal) and there is no obvious way of moving between both sections of the aircraft.

There was no provision for windows and the crew would use a rather complicated periscope system that would provide limited external views during take-off and landing. Price discounted visibility issues, saying that windows for the crew were largely unnecessary, as most of the flight would be automated. He also suggested that passengers would have nothing of interest to see at high altitude.

Four fully retractable struts would support the Lockheed saucer at rest and these might have been fitted with wheels to assist ground handling. In a typical operation, the aircraft

Above left: **Detail of the single-engine 'Island' unit for the Lockheed VTOL ramjet flying disc aircraft.** *US Patent Office*

Above centre: **Ducting system.** *US Patent Office*

Above right: **Side view of engine island unit.** *US Patent Office*

would take off vertically by directing the thrust from its engine island downwards. After lift-off, the engine's exhaust would be directed rearward and it would climb steadily to 10,000ft (3,048m) at a 10° angle. This would be progressively increased to 20°, until the vehicle reached a speed in excess of Mach 1 at 50,000ft (15,240m). The propulsive system would now be working in ramjet mode, with combustion taking place downstream from the engine module. Following air ducting changes, the engine(s) in the island would

Below: **An alternative engine 'Island' design for the Lockheed high performance ramjet-powered flying disc. This would use three gas turbines and, although more complex, would offer improved performance and reliability.**

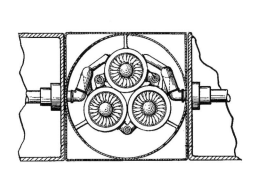

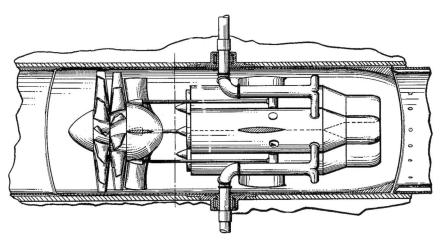

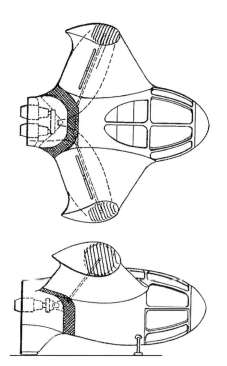

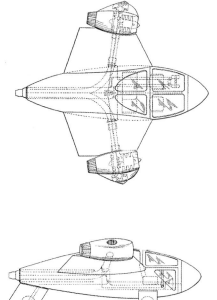

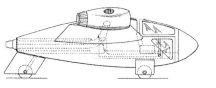

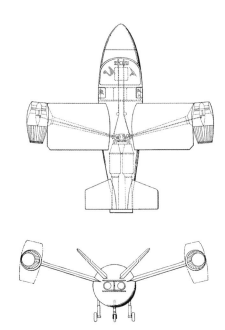

Two very unusual small VTOL utility aircraft designs produced by Nathan Price in the early 1950s. *Lockheed*

Above: **A compact VTOL aircraft design produced by Nathan Price in the late 1950s and part of a lengthy series of related studies.** *Lockheed*

operate at idle speed and the aircraft would reduce its angle of ascent to 10°, while continuing to accelerate until a maximum speed of Mach 4 had been reached at an altitude of 100,000ft (30,500m). An unrefuelled range of 7,500 miles (12,070km) was considered realistic for a military version, using hydrogen as fuel, with additional storage tanks carried.

There would have been numerous technical problems developing this interesting design and although most of the project's history remains unknown, it appears to have been abandoned in the late 1950s. Price also worked on a futuristic-looking four-seat VTOL utility aircraft powered by turbojets. He began work on this concept in the late 1940s and used jet engines for horizontal flight and wingtip pods containing shaft-driven turbines for lift. The design progressed into a small utility transport fitted with swivelling wingtip pods containing propulsive units and he described this as a vertically rising, road operable aircraft.

This gave way to a small oval-shaped aircraft using a dorsal turbojet in a nacelle that provided a full VTOL capability and thrust for horizontal flight. Intended for use as a high performance four-seat utility aircraft, it was fitted with two short downward-facing tail fins with control surfaces. The principal means of control during flight was vectored thrust; Price considered the option of attaching small wings

but decided it was unnecessary.

The aircraft would be equipped with a fully retractable undercarriage and the cockpit was integral with the forward fuselage section, providing good visibility. Although somewhat different in design, this concept may have been inspired by Alfred Loedding's previously described oval-shaped utility aircraft, which was published as a US patent on 25 November 1952.

This series of studies culminated in a very clean elliptical design, which was filed as a patent application (183108) in June 1956. Price's final wingless proposal featured improvements to cockpit visibility and may have been intended to attain supersonic speed in level flight. Conceivably, the previously mentioned designs were tested as wind tunnel models, but they are unlikely to have progressed much further.

LOCKHEED SKUNK WORKS FLYING CIGAR STUDIES

Two additional studies undertaken by Price during the 1950s are worthy of inclusion in this book as they evolved alongside the VTOL flying saucer design. Both of these proposals were for largely featureless 'flying cigars' with a similar appearance to some objects reported as UFOs.

The idea of developing an aircraft with no wings, no tail, no visible cockpit or obvious

control surface might seem rather extreme, but this is exactly what Nathan Price set out to do. He conceived a slightly flattened cylinder with an inlet at one end and an exhaust outlet at the other. This outwardly minimalist design was actually very complex and promised VTOL operation and very high performance. However, Price also mentioned the possibility of utilising features of this design in a circular or crescent-winged aircraft.

The turbo-ramjet-powered flying cigar would have high horizontal supersonic performance and a full VTOL capability. A forward-positioned cluster of turbojets would provide propulsion at lower speeds and the necessary lift for VTOL, with exhaust gases being channelled via upper vents to provide lift by means of the Coanda Effect. Several secondary turbojets would be located in the tail section to provide alternative propulsion in an emergency. Directional control would be achieved by means of vanes in the tail exhaust and the controlled release of gas through outlets along the fuselage.

Having attained supersonic speed, the turbojet engines would be shut down and the aircraft would fly on ramjet power. Price anticipated a maximum speed of Mach 4 and a ceiling of about 90-100,000ft (27-30,000m), with economic cruise at around Mach 3.5 at 75,000ft (22,860m). No figures were specified for range, although an intercontinental capability was suggested.

Flying Saucer Technology

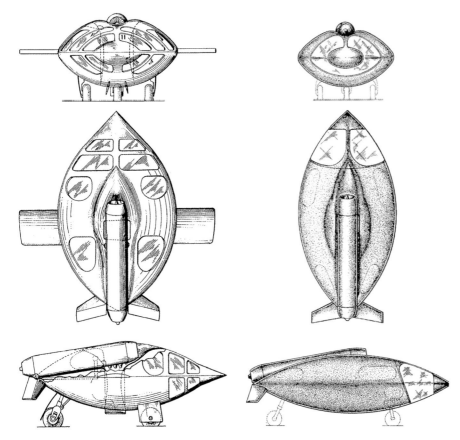

Left: **Two very advanced elliptical utility aircraft designed by Nathan Price in the late 1950s. The final version of this series is shown on the right.** *Lockheed*

The flying cigar was envisaged as a highly automated vehicle controlled by inertial guidance and programmed radar, although it was accepted that a crew was required to deal with problems or emergencies. Dimensions for this design are not quoted, but it is possible to deduce that the length was expected to be approximately 120-130ft (36-39m) and the width about 15ft (4.5m). The fully retractable undercarriage comprised four separate struts each fitted with four wheels.

This design was filed as a US patent on 13 August 1957, but was not published (ref: 3148848) until 15 September 1964 as a supersonic airliner. Several things point towards black budget origins of this project. First, the economics of developing a Mach 3.5 aircraft simply to carry 60 passengers makes almost no sense, especially in the mid-1950s. Then there is a seven-year delay before publication of the patent, and Nathan Price lists his residence in Mexico City. There is no mention of Lockheed on these documents,

although they do appear on some related developments.

About six months after filing the first patent an extensively revised design was submitted as a patent, which would be published several years later (ref: 2973921). This concept differed considerably from the original design and employed a ring of 12 turbojets at the front of the aircraft. Price also produced a more aerodynamically efficient fuselage. Other differences include a revised VTOL ascent and descent efflux nozzle system. Price acknowledged the probability of noise issues within the aircraft and suggested measures to reduce this. Surprisingly, the passenger capability appears to have been almost halved for this version, which makes even less sense than the earlier proposal. Again, there is no mention of Lockheed in the documentation and Price is now shown as living in Geneva, Switzerland.

For what it is worth, this author is guessing that these designs started life as a strategic cruise missile that was derived from the flying saucer. It was originally intended to carry a thermonuclear warhead, but Price and his Skunk Works team realised there was potential for manned applications. Conceivably there were tests undertaken involving wind tunnel models, but it seems unlikely that Price or Lockheed ever seriously anticipated any civil applications for this flying cigar aircraft.

WIBAULT'S VTOL CONCEPTS

Michel Henri Marie Joseph Wibault (1897-1963) was a well-respected French aeronautical engineer with many successful aircraft designs to his credit.

During the immediate post-war years there was considerable European interest in military VTOL aircraft, which would lead to some extraordinary projects such as the French SNECMA tail-sitters derived from advanced wartime German research. Wibault hoped to capitalise on this perceived need for fast aircraft capable of operating from improvised sites and by 1950, he had started working on the design of a flying saucer called the Gyropter.

This one-man design was based on a

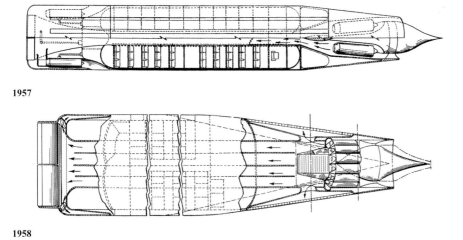

1957

1958

Left: **Developed alongside the disc-shaped VTOL ramjet aircraft by Nathan Price's design team, these very unusual Lockheed wingless VTOL proposals were probably conceived for military use.** *US Patent Office*

largely enclosed rotor system driven by four tip-mounted turbojet or ramjet engines protruding from the fuselage underside. This would provide full VTOL, hover capability and horizontal flight while also largely eliminating torque problems and generating gyroscopic stability. Air would be drawn in around the upper stator and driven downwards, while fuel (and possibly air) would be fed along each hollow rotor arm. Forward flight would be achieved by adjusting the airflow from the rotor, and the initial Gyropter concept had four ballast tanks located around the rim of the disc for controlling trim, plus an upper stabilising fin and rudder. The cockpit was centrally located and covered by a dome-shaped transparency, and the aircraft would be supported on the ground by three or four fully retractable struts with wheels.

No details of dimensions, weights or performance are available for the first Gyropter, although assuming it had been built, the aircraft would have been approximately the same size as the Avro-Canada Project Y2. Wibault followed this design with a more efficient-looking triangular or deltoid aircraft that used the same rotor system. It was equipped with a single rear-mounted jet engine and the cockpit was moved forward to improve

visibility. Large control surfaces were fitted to the trailing edge, plus a tail fin and rudder. This version of the Gyropter would also be equipped with a fully retractable undercarriage.

The triangular Gyropter had more potential for development than the flying disc and Wibault evolved this concept into an even larger spade-shaped aircraft equipped with three enclosed rotors and three turbojets in the tail. These Gyropter designs were completed in 1953 and patents were filed by the Vibrane Corporation in New York City.

The design continued to evolve, with the aircraft becoming more conventional in appearance and Wibault moving to a new type of propulsive blower system. He now envisaged a new single-seat VTOL tactical nuclear strike aircraft that could operate from an improvised site. The propulsion system took the form of four blowers positioned around the aircraft's centre of gravity. Powered by a Bristol BE.25 Orion turboshaft engine producing 8,000hp (5,970kW), each blower would rotate to produce downward or horizontal thrust.

In 1955, Wibault approached the French government with proposals for a new aircraft, but his ideas were rejected. He then went directly to the NATO Mutual Weapons

Development Programme (MWDP) Office in Paris who forwarded his studies to Theodore von Karman for review. These clearly impressed von Karman, who then passed them to Stanley Hooker at Bristol Aero Engines in England. Subsequently, a meeting was arranged and Hooker assigned Gordon Lewis, Pierre Young, and Neville Quinn to examine Wibault's ideas in detail. It soon became apparent that this propulsion system had great potential but could be significantly improved by directly channelling the engine's exhaust via swivelling ducts. This led to the construction of a new engine called the BE.48. This Orion-based system soon developed into the BE.52 and then the BE.53 based on the Orpheus turbojet with a larger compressor fan.

Wibault had been retained by Hooker as a consultant and is said to have been very pleased with the direction this project was taking. In January 1957, Bristol-Siddeley Engines Ltd applied for a UK patent on behalf of Wibault and Lewis for the new engine and although Bristol now had a new type of advanced propulsion unit in development, there was no suitable aircraft to utilise it.

Below: **The original circular VTOL Wibault aircraft design (left) and a more advanced triangular concept. Both use the same internal rotor system.** *Bill Rose*

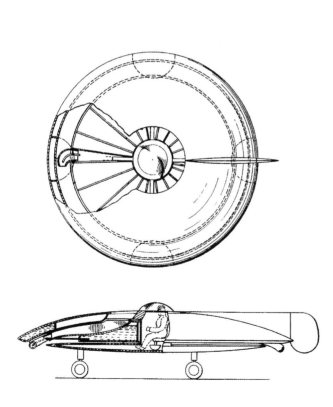
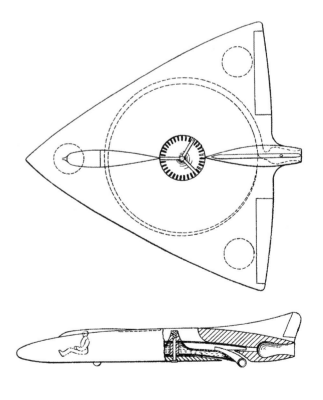

Below **Two variations of Wibault's tip-driven rotor design.** *Bill Rose*

Right: **An advanced Wibault three-rotor VTOL transport aircraft design.** *Bill Rose*

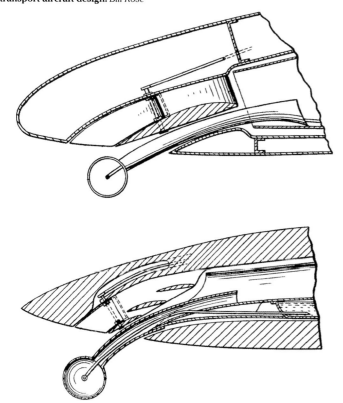

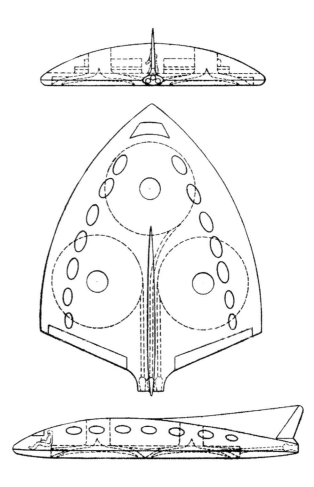

Fortunately, Hawker Aircraft's chief designer Sydney Camm had taken a keen interest in the engine project and assigned a team to develop a suitable airframe. This led to the Hawker P.1127 and eventually the Harrier, which would become the world's first successful VTOL combat aircraft.

COUZINET'S FLYING SAUCER

René Alexandre Arthur Couzinet (1904-1956) was a French aviation pioneer and designer who funded development of a manned aerodyne vehicle during the 1950s. The son of a schoolteacher, Couzinet grew up in Saint-Martin-des-Noyers, and as a child he developed an interest in the way birds flew and in manned flight. In 1924, he joined the Higher School of Aeronautics and by this time had filed several aviation patents, including plans for a helicopter.

In 1925, Couzinet became a junior Air Force officer and following the historic transatlantic Lindbergh flight of 1927, he was given the task of developing a French aircraft with an intercontinental capability. This led to a three-

engined design called 'Arc-en-Ciel', which allowed Couzinet, his close friend Jean Mermoz and five crew members to make the first successful double crossing of the South Atlantic in 1933. In December 1936, Mermoz was killed in a flying accident and Couzinet was deeply

Above: **René Alexandre Arthur Couzinet.** *Bill Rose*

affected by this, although he married Mermoz's widow Gilberte Chazottes in 1939.

At the start of the war in Europe, Couzinet and his wife moved to Brazil. Couzinet was appointed as Director of National Aircraft Production and he taught aeronautics at the Brazilian Air Force Training School. However, it became impossible to buy aircraft components from France which was now under Nazi rule and equally difficult to do business with the Americans after they entered the war. For the remainder of World War 2, Couzinet's organisation was involved with the manufacture of gas generators and engineering equipment, although he was able to continue with aircraft design work.

After the war, Couzinet became interested in reports of UFOs and flying saucers, turning his attention to developing a disc-shaped aircraft with VTOL performance. Having returned to France, he filed several patents during May 1955 for aerodyne flying disc designs. These included drawings and details of a demonstrator and more advanced designs for single- and twin-seat aircraft.

By autumn 1955, engineers working at the

The Post-War Era

109

Levallois-Perret factory had completed a largely functional mock-up of the Aerodyne Demonstrator, which was revealed to the media. It had an overall diameter of 29ft (8.84m), with a centrally located cockpit covered by a dome-shaped transparency. A fixed position tricycle undercarriage provided support, and this would be fully retractable on a flightworthy version.

Lift for VTOL was generated by a partly shrouded dual rotor system fitted with ninety-six peripheral blades that were driven by three 180hp (134kW) piston engines. The rotation was also expected to generate gyroscopic stability, and a single low power Turboméca Palas turbojet, housed in a ventral pod, would be used for horizontal flight. Flight control took the form of a flap behind the cockpit, a vertical control surface behind the jet engine, which was also designed to swivel.

Couzinet maintained that this aerodyne

Above: **An edge-on view of the Couzinet Aerodyne engineering mock-up. The rotor blades appear to functioning in this image.** *Bill Rose*

Below: **The Couzinet Aerodyne engineering mock-up. This image shows the rotor blades in motion, revolving around the circumference of the disc.** *Bill Rose*

Flying Saucer Technology

Right: **Rotor detail of the Couzinet Aerodyne engineering mock-up.** *Bill Rose*

was simply an engineering mock-up, but it has been said that although built mainly from wood, the vehicle was fairly close to a functional prototype. Two Lycoming piston engines had been installed to drive the metal rotors and during static tests runs they turned at 80rpm. It remains unclear if there were serious intentions to build an airworthy version of this demonstrator, as Couzinet hoped to progress directly to a full-sized prototype called RC.360. This was already in the early stages of development at the Roche-sur-Yon factory. Models had been wind tunnel tested and reports suggested that the first test flight would take place in April 1956.

The RC.360 had a diameter of 44ft 7in (13.6m) and a lifting surface area of 645.6sq ft (60.0m2). It utilised the same principle for lift shown on the demonstrator, but each of the larger rotors would have 50 adjustable vanes instead of 48 and they would be driven by six Lycoming engines instead of three. For horizontal flight, the RC.360 would initially be equipped with a licence-built Armstrong-Siddeley Viper turbojet, delivering 1,639lb (7.2kN) of static thrust.

In flight, the aircraft would be controlled by movable surfaces on four upper and lower swept fins. It is probable that additional control surfaces were contemplated. Several

positions were considered for the jet engine that might have been housed in a fixed nacelle below the aircraft or a pod above the fuselage. Various cockpit arrangements were also proposed, including accommodation for a single pilot, a side-by-side arrangement for two crew members and two separate cockpits. Couzinet was also aware of visibility issues and proposed downward viewing ports as necessary. Gross weight of the prototype RC.360 was set at 27,700lb (12,565kg). Maximum speed, ceiling and range have never been quoted, but are unlikely to have been very impressive.

A further development of this concept was an amalgamation of the circular wing with a conventional tubular fuselage, although this idea did not progress beyond the initial stages. The role Couzinet envisaged for his aerodyne is not entirely clear, although available documentation suggests that he visualised the RC.360 as a military aircraft able to operate from improvised sites.

Nevertheless, it is hard to imagine this design providing the flexibility of a conventional helicopter and it was never going to compete effectively with the supersonic jet fighters that were already in development.

Once initial interest in the flying saucer subsided, it became clear there was no commercial interest in the project. Unable to raise sufficient money to complete the RC.360, Couzinet sank into deep depression, and on 16 December 1956, he and his wife Gilberte committed suicide in their apartment. It was a sad ending for a gifted aviation pioneer.

The aerodyne mock-up vanished without

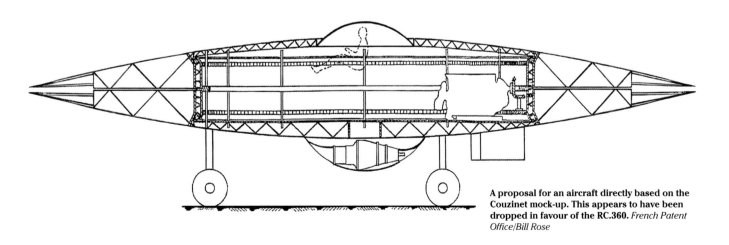

A proposal for an aircraft directly based on the Couzinet mock-up. This appears to have been dropped in favour of the RC.360. *French Patent Office/Bill Rose*

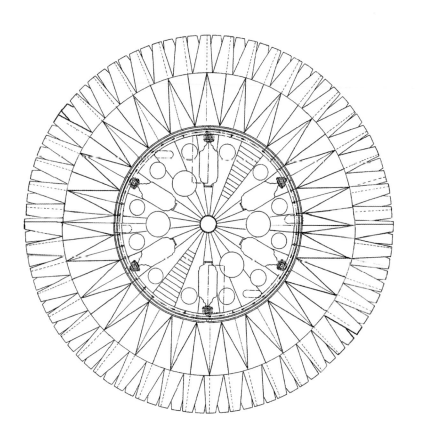

Rotor detail for the Couzinet RC.360.
French Patent Office

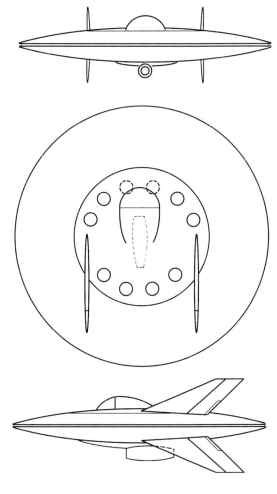

Initial design for the Couzinet RC.360 Aerodyne with side-by-side cockpit. *Bill Rose*

A wooden scale model of the two-cockpit Couzinet RC.360 flying disc with overhead engine pod. *Bill Rose*

Flying Saucer Technology

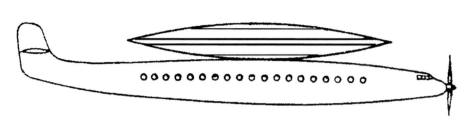

a trace, although it was probably broken up for scrap. Nothing is currently known about the documentation for this project, which may be lost. Ironically, Couzinet is now recognised in France for his outstanding contribution to aviation and his name appears on postage stamps, roads, a college and a rural airport.

PAYEN

Nicolas Roland Payen was another French designer who became fascinated by the behaviour of dart-shaped paper planes when he was a child. Payen completed his first aircraft proposal by the age of 14 and became a leading proponent of the delta wing, with some supporters claiming that he conceived the idea before Alexander Lippisch.

Payen entered the world of aviation engineering in the early 1930s, soon becoming a prolific designer. He created plans for hundreds of different aircraft during his career, but today is mainly remembered for his unusual and rather futuristic concepts which still appear visually striking and often rather quirky. A typical Payen aircraft would utilise a rear-mounted delta wing with the tail forming part of the fuselage and small, forward-mounted canard wings. In fact, some of the credit for early designs using this unusual layout belong to Payen's friend and fellow designer Robert Sauvage who filed a French patent (729,568) with him on 26 April 1932 for an aircraft of exactly this type.

After World War 2, Payen set about designing a jet fighter for the Armée de l'Air (French Air Force), which was called the Payen P.48. His proposal was rejected, seemingly on the grounds of Payen's company having insufficient means to develop the aircraft. However, in 1951, he made use of his Pa.48 design to form the basis of a small experimental jet aircraft called Pa.49A.

Construction was soon under way, using metal in essential areas and wood/plywood

wherever possible. Payen received some official assistance with this demonstrator and wind tunnel testing began in 1952. By mid-1953, a single Turboméca Palas turbojet had been installed with air inlets positioned at the leading edge wing roots. Taxiing trials began at Payen's factory and the first test flight took place on 27 November 1953, with Pa.49A piloted by Antoine (Tony) Ochsenbein. This was the first French jet-powered delta wing aircraft to fly, although officially the initial flight took place on 16 December 1953.

The delta winged Pa.49A was fitted with a single-seat cockpit faired into the tail fin. Its wingspan was 16ft 11in (5.15m), with a wing area of 123.7sq ft (11.5m²), a length of 16ft 9in (5.10m) and a height of 7ft 6in (2.3m). Despite the engine's relatively low performance, Pa.49B achieved a maximum speed of 248mph (400kph) and its ceiling was 26,000ft (8,000m). Gross weight was approximately 1,733lb (647kg) and the tricycle undercarriage was fixed in position, with the fairings around each strut to improve streamlining.

Test flights of the diminutive jet-powered delta continued during 1954, with further modifications taking place, leading to the aircraft being redesignated Pa.49B and given

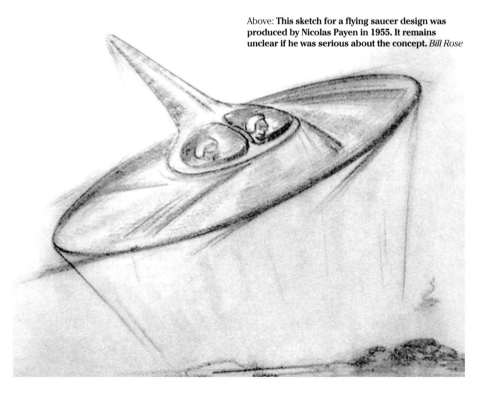

Above: **This sketch for a flying saucer design was produced by Nicolas Payen in 1955. It remains unclear if he was serious about the concept.** *Bill Rose*

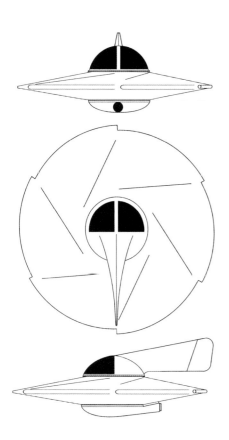

Above: **General appearance of the Payen flying saucer proposed in 1955.** *Bill Rose*

the name Katy. Test flights continued into 1955, when the Pa.49B Katy was demonstrated at the Paris Air Show. It is said that Payen hoped to follow the Pa.49B with a larger jet-powered aircraft but the project ended in 1956. Many years later, the Pa.49B prototype was passed to the Aviation Museum at Le Bourget, where it is now on display.

While the Pa.49B was undergoing trials, Payen became interested in the idea of building a circular-winged aircraft, un-doubtedly as a consequence of devel-opments taking place at René Couzinet's factory. By the end of 1955, Payen had started to sketch ideas for a small flying saucer aircraft which he designated Pa.61 (although some references rather confusingly quote this as Pa.64).

Few details of the Pa.61 exist, but it looks as if this design was something of an outgrowth of the Pa.49B, perhaps intended to compete with Couzinet's aerodyne. Slightly larger than the Pa.49B, this Payen concept has a centrally located cockpit with accommodation for two, and Payen probably envisaged the aircraft in a basic utility role. It would have a rotating wing section forming the outer part of the fuselage and a large stabilising fin attached to the rear of the cockpit.

The lower section of the aircraft beneath the cockpit would have housed a Turboméca Marboré II jet engine for horizontal flight, although it is not clear if this engine was also

expected to drive the rotating wing, as the Marboré II produced a maximum static thrust of only 880lb (3.9kN). Support on the ground is also unclear and a retractable underc-arriage could only have been accom-modated in the lower central section, which was mostly occupied by the engine. Needless to say, this idea did not progress much further than the elementary design stage.

ASTRO KINETICS VTOL AIRCRAFT

At the beginning of the 1960s, a small group of engineers based in Houston, Texas, decided to establish a company called the Astro Kinetics Corp. Their aim was to build a small one-man flying saucer demonstrator that would lead to larger commercial vehicles using the same design principles. Heading the company was Fremont W. Burger; his chief engineer was James W. Miller, assisted by Harvey H. Houde.

A small one-man prototype was completed in early 1963, which used an impeller driven by a 135hp (100kW) Mercury outboard motor in the centre of a large upturned dish. This would create an airflow over the upper surface and generate lift. The pilot and engine were carried in an open framework below the dish, and the aircraft was supported on the ground by three fixed landing legs. Little is known about the

The first VTOL prototype produced by Astro Kinetics of Houston in 1963. The design appears to have been abandoned soon after completion. *Astro Kinetics Corp*

Flying Saucer Technology

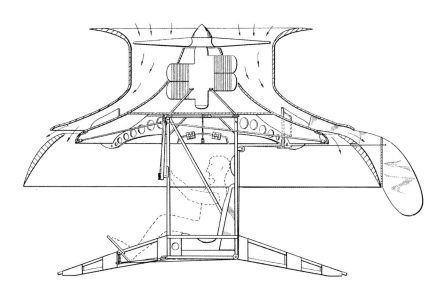

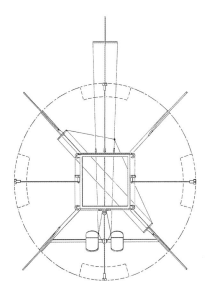

Above: **US patent drawing for the second Astro Kinetics one-man VTOL craft.** *US Patent Office*

Right: **Initial design for the second Astro Kinetics one-man VTOL platform, which was produced in 1964.** *US Patent Office*

Below: **A press demonstration of the second Astro Kinetics Dynafan VTOL craft in the company's Houston hangar. This is believed to have taken place in December 1964.** *Astro Kinetics Corp*

anticipated performance of the Astro Kinetics Mk 1 VTOL demonstrator. It was expected to function like a small autogyro, but may not have performed as planned during tests.

A second version was completed towards the end of 1964 and this appears to have been a completely redesigned vehicle. Astro Kinetics then applied for a US patent, which was published in October 1966 (3276723). This version was called the Dynafan and used a two-blade fan driven by a Chevrolet Corvair turbocharged 148hp (110 kW) engine. Lift was achieved by directing air downwards through a large duct and out over the airfoil below. Directional control would be undertaken using internal control of the airflow using flaps and a small tail rudder.

In this design, the pilot was located in a framework below the dish and supported on the ground by three rigid legs. The upper circular components were made from fibreglass and the lower framework was fabricated from steel and aluminium. The engine was supported directly above the pilot within the centre of the dish section. Overall diameter of the Dynafan was 7ft 7in (2.3m) and the height was 8ft (2.4m). The prototype weighed 680lb (308kg) and endurance was estimated at two hours.

In December 1964, Astro Kinetics invited the press to view the Dynafan undergoing a tethered hover demonstration at the company's hangar. James Miller, who had already completed two hours of similar trials, flew the prototype. Nothing came of this design and there is no evidence to show that the Dynafan ever became airborne.

Company president Fremont W. Burger had intended to show that the Dynafan could be handled as easily as an automobile and he suggest that a future aircraft based on the Dynafan might have a diameter in excess of 100ft (30.4m) and be capable of lifting a 250,000lb (113,398kg) payload. However, there was no financial backing for the Dynafan, which seems to have been abandoned during the following year. Burger died in June 1965, aged 49, and this may have had some bearing on events.

THE VOUGHT ADAM PROJECT

During the 1950s, a number of US aircraft manufacturers explored ideas to improve STOL performance and in some cases provide a VTOL capability. One contractor developing this type of technology was Chance Vought Corp, which started work on a system called Air Deflection and Modulation (ADAM) in 1958. This would eventually produce a series of designs that some observers have come to regard as obvious successors to the Zimmerman XF5U-1 Skimmer.

The initial aim of Vought's engineers was to provide enhanced STOL performance for light transport aircraft using wing-mounted propulsive pods and exhaust deflector flaps. This led to the construction of a one-quarter scale model of the system, with support from NASA Langley and NASA Ames. The design evolved into a propulsive system contained within the wing, with the exhaust flow being vectored downwards for VTOL and brief hover, using louvred shutters. Two gas generators coupled to the fans would be housed in nacelles on each side of the fuselage.

In 1963, Vought received funding from the USAF and US Army to investigate military

applications for ADAM, which extended from reconnaissance/strike aircraft to light transporters. Soon after the start of wind tunnel testing it was found that the ADAM concepts would require substantial tails to address centre of gravity issues. It was then decided to install a fan in the nose to deal with pitch problems during hover and, as a consequence, the tail size was reduced.

Senior staff at the Dallas Design Office were convinced that it would be possible to build a VTOL strike aircraft with a good payload capacity that possessed the ability to reach transonic or supersonic speeds in level flight. At that time, Vought were still in the process of reforming into Ling-Temco-Vought (LTV) and the ADAM project was felt to be of considerable importance to the company's future success. Subsequently, LTV added additional financial backing to the programme and anticipated the construction of a proof-of-concept demonstrator. LTV executives also felt certain that the ADAM strike aircraft would win favour with the US Navy and this led to a campaign of promotion and lobbying.

By this time many different concepts had been produced, including transport aircraft,

The two-seat Vought ADAM I basic configuration proposed in November 1962. Note the twin tail fins. *Ling-Temco-Vought/Bill Rose*

although the focus of attention was securing contracts for a VTOL fighter or strike /reconnaissance, close-support aircraft.

ADAM I was the basic model that emerged in 1959 and this design rapidly evolved into ADAM II. Numerous improvements were made to the design and layout, with the gas generators being shifted into the fuselage and reduced from four to two. It was anticipated that the aircraft would remain flyable in the event of one engine experiencing sudden failure.

A fan in the nose of the aircraft was used to direct air downwards during VTOL and hover. It would be shaft coupled to the engines and provide pitch stability, although at the expense of a radar system. The ADAM II combat aircraft was supported on the ground by a fully retractable tricycle undercarriage, with the main supports and wheels housed in the booms. This would later be revised into a bicycle configuration with outriggers. Small stabilising winglets, a tail and rudder were fitted, while the deflector flaps would take on the function of elevons in level flight.

Above: **Model of an LTV ADAM II basic con-figuration aircraft equipped with hardpoints carrying bombs. The nose intake for the forward fan and wing fans are clearly visible in this photograph dating from about 1967.** *Ling-Temco-Vought*

Left: **An ADAM model with twin tail fins is supported upside down in the LTV low-speed wind tunnel.** *Ling-Temco-Vought*

The single-seat ADAM II strike aircraft would have a gross weight 17,400lb (7,892.5kg) and a two-seat variant was proposed with more powerful engines and an anticipated gross weight of approximately 30,000lb (13,608kg). A supersonic ADAM fighter was still considered feasible, along with a 75-ton civil and military S/VTOL transporter.

However, the USAF and US Army lost interest in ADAM during 1965 and withdrew further financial assistance. The British Harrier VTOL fighter was under review and conceivably this had some bearing on the issue. Nevertheless, LTV continued to develop the ADAM concept and they still hoped to win orders from the US Navy or Marine Corps.

The final incarnation of this series was the

Above: **Artwork showing a USAF ADAM II aircraft operating in the close-support role. It was anticipated that ADAM II aircraft would start to enter service in the late 1970s.** *Ling-Temco-Vought*

Below: **The most advanced design to emerge from the ADAM project was the ADAM III proposal of June 1968. Streamlining was improved and the fan had now been relocated behind the cockpit, allowing the carriage of a radar system.** *Ling-Temco-Vought*

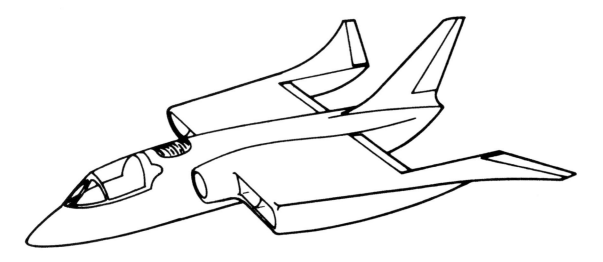

Flying Saucer Technology

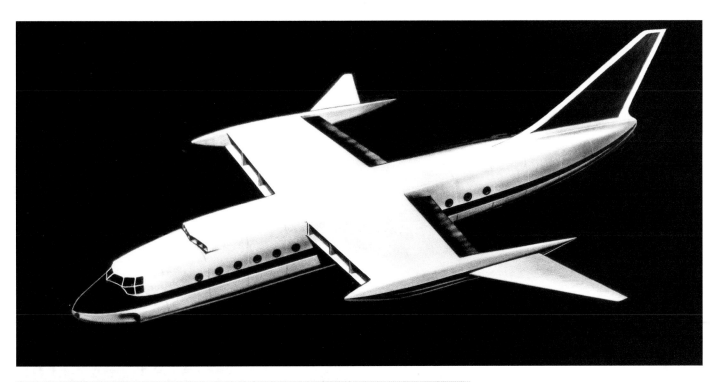

Top: **Artwork showing a single-seat Vought ADAM II aircraft in clean condition from February 1967. From the conception of this design there were numerous changes that included relocation of the engines, alterations to tail fins and modifications to the forward lift fan, seen with cone positioned in the nose inlet.** *Ling-Temco-Vought*

Above: **One of many possible applications of the ADAM system to civil and military transport aircraft. This artwork shows a 75-ton proposal.** *Ling-Temco-Vought*

ADAM III strike aircraft. It was sleeker, carried both engines in wing root nacelles and the nose fan was moved to a position behind the cockpit, allowing the use of airborne radar. Work on ADAM III and related S/VTOL designs continued until late 1968 when LTV appear to have accepted that there was no possibility of changing minds at the Pentagon. ADAM seems to have been a promising project, which makes it hard to understand why the military decided not to proceed.

MOLLER'S FLYING SAUCERS

Paul S. Moller (1935-) was born in British Columbia, Canada, and from an early age showed a natural aptitude towards design and engineering. Having graduated from High School, his interest in aviation led him to enrol at Trade School for a three-year aircraft maintenance course. He then became an engineer with the Canadair Aircraft Company based at Montreal and during this period he was accepted as a student by McGill University. Having graduated with a Masters in Engineering and then a PhD, Moller was offered a position with the University of California (UC Davis) in 1963 as a Professor of Mechanical Engineering. Having accepted, he remained in this post for the next 11 years, spending much of his spare time working on a VTOL aircraft project which had started life as a small-scale model, built in Canada. Having set up the Moller Aircraft Company, he began to assemble a manned craft in his garage, which received the designation XM-2.

This one-man saucer-shaped design had an overall diameter of 14ft (4.27m), a gross weight of 700lb (317kg) and was powered by two ducted fans driven by two-stroke McCulloch engines. Directional control was attained by flaps on the underside of the craft. It remains unclear if Moller had any realistic expectation of achieving flight, initial tests undertaken in 1965 showing that the engines were incapable of lifting XM-2 out of ground effect. As a consequence, Moller uprated the XM-2's engines, thanks in part to a UC Davis grant, and also filed for a US patent (3410507) describing XM-2 and a more advanced design

Above: **The XM-2 VTOL vehicle built by Paul Moller during the mid-1960s in his garage. It lacked sufficient engine power to lift out of ground effect and is said to have been unstable in operation.**
Moller International

Below: **General appearance of the Moller XM-3 built as a prototype in the late 1960s.**
US Patent Office

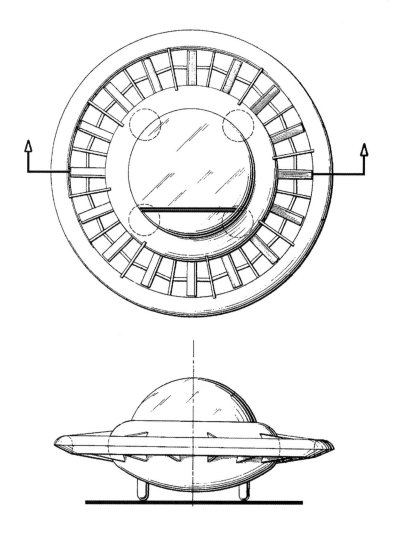

which would become the follow-on model.

In 1966, Moller arranged a public demonstration of XM-2 at UC Davis Airport, but the newly installed Mercury outboard engines provided little improvement over the original engines and he was unable to rise much higher than about 3ft (0.91m) from the ground. XM-2 proved very difficult to control and the project was soon abandoned, with Moller deciding to start work on a new vehicle called XM-3.

This differed considerably from XM-2, having the appearance of a compact one-man machine straight out of a SF comic book. Propulsion was achieved using a ring fan driven by eight small air-cooled go-kart engines, and most of the vehicle was built from fibreglass. XM-3 was completed in 1968 and Moller attained a height of about 10ft (3m) during the first test flight, while also finding it possible to manoeuvre fairly easily. Although the design was less than perfect, Moller was now encouraged to establish the M Research Company. In 1969, M Research obtained a number of Wankel rotary engines from West Germany and began adapting them for use in a new Moller design known as the XM-4. Moller's business had now diversified into the manufacture of exhaust systems for motorcycles and this provided the funding for M Research to become the Discojet Corporation in 1971, which set up at new business premises in Davis, California.

XM-4 had been in development since 1969 and details of the design were filed as a US patent (3614030). The prototype was largely completed by 1973 and this design was significantly more sophisticated than the previous two prototypes, with accommodation for two persons. The small saucer-shaped aircraft had a diameter of 9ft 3in (2.8m) and an overall height of 3ft 4in (1m). Eight separate ducted fans, each powered by a 300cc Fichtel-Sachs Wankel rotary engine, provided propulsion. Each ducted fan unit was also designed to tilt between 2 and 4° from the central axis, providing directional control.

Initially, Moller claimed that that calculations based on engineering studies and wind tunnel tests would allow full VTOL operation, a maximum speed of 200mph (320kph) a ceiling of 15,500ft (4,724m) and a range of 325 miles (523km) carrying a payload of 1000lb (453kg), later revised to less than half this figure. However, once hover trials began in 1974, it became evident that XM-4 lacked sufficient power to reach a significant height and that the original claims were wildly unrealistic.

Flying Saucer Technology

FILIMONOV'S HYBRID

In the early 1980s, a Russian engineer called Alexander Filimonov began to consider ideas for a new aircraft that drew on concepts produced in America by Naught and Ross. This led to the development of a hybrid fixed wing aircraft with a circular central body that integrated features of the helicopter, hovercraft and airship. His objective was to create a transport aircraft capable of operating in difficult environments, taking off and landing on rough ground, snow or water. Filimonov hoped to fill an existing gap and provide the means of transporting cargo and passengers to remote hard-to-access outlying locations such as oilfields or sections of pipelines requiring repair. Other possible uses included firefighting, medevac and specialised military operations.

Filimonov worked at the Tyumen Industrial Institute and by 1987 had secured considerable support and State funding to develop and build a prototype called Bella-1. Wind tunnel tests of models began in 1989 and the prototype was completed in 1994. A company called Tumenecotrans was formed to support the aircraft, with assistance from the Moscow Aviation Institute, Siberian Scientific Research Institute of Aviation and Aircraft Repair Factory No 26 at Tyumen. Ground tests of Bella-1 began at the Ulianovsky Avia Industrial Complex and in 1995 an application was submitted for a Russian Federation patent, which was published in 1997 (2092381).

The Bella-1 prototype was designed to be powered by American-built engines as no suitable units were available within the CIS. Two were mounted externally behind the central disc section in a pusher configuration. A third separate piston engine would drive the rotor in the duct to provide lift. Shutters were fitted above the duct to regulate airflow and the lower area of the duct was surrounded by a hovercraft-style skirt to provide an air cushion. The aircraft was equipped with skis fitted with wheels allowing operations on a range of different surfaces and it would be lightened by the use of helium in tanks within the central section. Bella-1 was fitted with a boom tailplane linked by a horizontal stabiliser. Two rudders

In 1981, the Discojet Corporation changed its name to the Moller Corporation and the XM-4 remained an ongoing project, although work on the prototype had slowed to a snail's pace. Eventually, in 1987, the machine was refitted with more powerful and lighter weight Wankel rotary engines provided by Outboard Marine Corp (OMC) and XM-4 was redesignated M200X. Trials were under way by 1989, although the M200X still failed to achieve the original design goals set out for the XM-4 some twenty years earlier.

Dr Moller followed M200X with a more ambitious prototype called the M400 Skycar. This automobile-sized vehicle used four vectorable ducted fan units with the anticipated capability of cruising at 350mph (563km/h) and attaining an altitude in excess of 30,000ft (9,144m), with an unrefuelled range of around 900 miles (1,448km). Like the earlier prototypes, M400 never came close to achieving its design aims and the future of Skycar is rather doubtful.

Rear view of the Tumenecotrans Bella-1 during ground testing trials. *Tumenecotrans*

The Tumenecotrans Bella-1 approaching lift-off during winter trials. *Tumenecotrans*

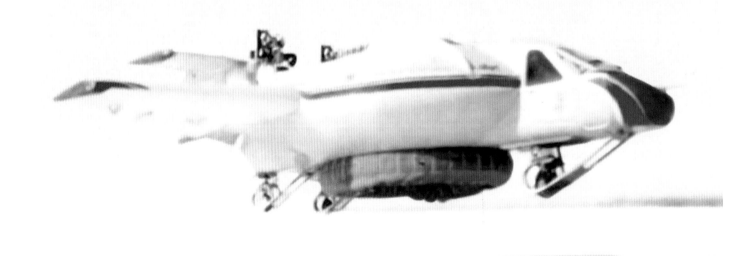

The Russian-built Tumenecotrans Bella-1.
Tumenecotrans

Believed to be currently mothballed, the hybrid
Tumenecotrans Bella-1 in a small hangar.
Tumenecotrans

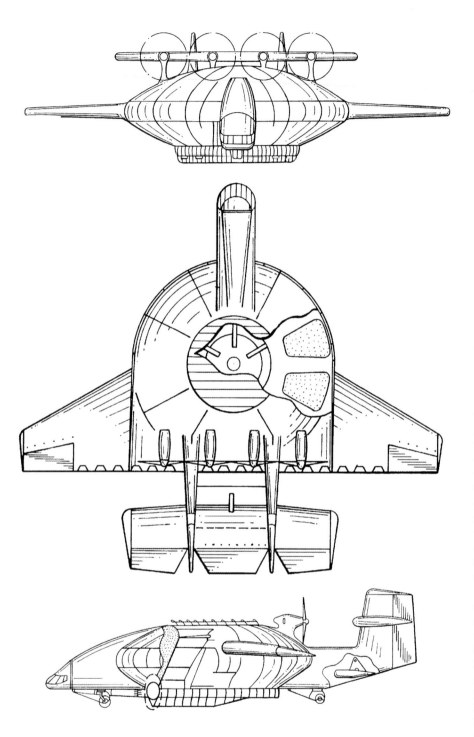

Above: **It was hoped that the hybrid Tumenecotrans Bella-1 aircraft would lead to a much larger multi-engined transport aircraft with a gross weight of about 350 metric tonnes. With a long range and the ability to reach inaccessible areas, this design was expected to attract civil operators and the military. However, economic pressures brought State support for the project to an end and it was abandoned.** *Russian Agency for Patents & Trademarks*

and elevons in the wings provided flight control. It appears that Filimonov originally hoped for a VTOL capability but accepted he would achieve no more than good STOL performance.

smothballed.

As a footnote, I should mention that this aircraft is not a winged version of the Aviastar Thermoplane, as incorrectly suggested on some internet sites.

EKIP

Described by the international press as Russia's UFO, the EKIP ('Ekologiya i Progres' – Ecology and Progress) was an unusual experimental aircraft developed by Saratov Aviation during the early 1990s. EKIP was conceived by an engineering team working under the direction of Professor Lev Nikolaevich Schukin and Alexander Sobko.

Aside from its unusual lifting body shape, EKIP employed a number of very interesting features that included a largely unproven vortex oscillating propulsion system and a method of boundary layer control. When fully developed, the production versions of EKIP would be supported on a cushion of air and be able to fly just above the ground or water, reaching a maximum speed of about 100mph (160kph). EKIP was also designed to fly like a conventional aircraft, with a maximum speed of 435mph (700kph) and a ceiling of about 42,000ft (12,800m). The compact twin-engined, manned version would have a payload capability in excess of 16 metric tonnes and the large multi-engined transporter would be able to carry 120 metric tonnes over a range of 3,278 miles (6,000km).

A cushioned landing support system would allow EKIP to operate from most ground surfaces or fairly smooth water, but this aircraft was not designed to have a VTOL capability and could not hover in the air. The twin-engined EKIP was expected to be able to take off in a distance of about 1,475-1,640ft (450-500m) and the transport version would require approximately 1,968ft (600m). It would also be possible to make very steep descents at an angle of attack as great as 40° and Schukin claimed that EKIP could make an emergency landing without engine power.

The initial phase concentrated on development of the unique propulsion system, with some of the research being conducted by specialist organisations. After a series of models were wind tunnel tested, Saratov built a small remote control demonstrator designated L2-1, which flew for the first time in 1994. This was followed by a larger single-engined unmanned aircraft known as L2-2, which was apparently built by S. P. Korolev RSC Energia and assembled at Saratov Aviation's facility. There were stability problems with both model demonstrators that led to control fin modifications, but the trials went reasonably well and it was decided to construct an even larger unmanned prototype called the L2-3.

This aircraft had a wingspan of 61ft 2in (18.64m), a length of 37ft 2in (11.33m) and a height of 12ft 3in (3.73m). The calculated maximum speed would be 379mph

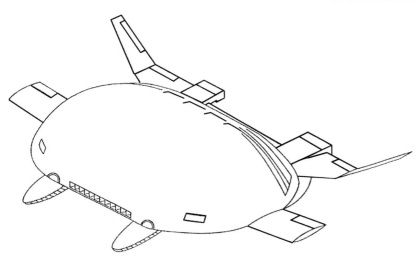

(610kph) with a take-off requirement of 1,476ft (450m), ceiling of 38,000ft (11,500m) and a range of 1,550 miles (2,500km). L2-3 had a gross weight 26,445lb (12,000kg) allowing a payload of 8,818lb (4,000kg). It is believed that this prototype was designed to accept the installation of a cockpit and ejector seat if required. Power for horizontal flight was provided by two turbofan engines with an afterburning system to improve the short take-off capability. Air inlets fothese engines would be positioned at the aircraft's leading edge and their exhaust would be directed through letterbox-shaped outlets capable of thrust vectoring. In addition, two separate gas turbines would be used to provide lift via a system of louvred shutters beneath the aircraft and all four engines would be capable of running on a variety of different fuels from kerosene to propane. L2-3 utilised a steel airframe, with composite materials used wherever possible.

The civil and military potential for this design seemed significant and EKIP was considered relatively easy to build, expected to be quiet in flight and economic to operate. But there was no government funding available for experimental designs like EKIP and the inability to generate commercial interest meant that this project was mothballed in the late 1990s.

EKIP was effectively finished, but then the US Navy's Naval Air Systems Command (NAVAIR) took an interest in the project. They thought the design showed considerable promise and by early 2005, there were reports in the press that the US Navy had signed a development contract with Saratov Aviation. Now there were suggestions that a new unmanned version of EKIP weighing 500lb (226kg) would fly at the Patuxent River Naval Air Station in 2007.

No further details of this joint project were released and when the author contacted NAVAIR in 2006, it was made clear to him that EKIP had been cancelled and there were no plans to pursue further development. NAVAIR were unwilling to expand on their reasons for dropping EKIP and it remains unknown if there were unexpected technical issues,

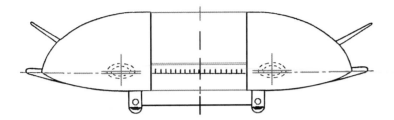

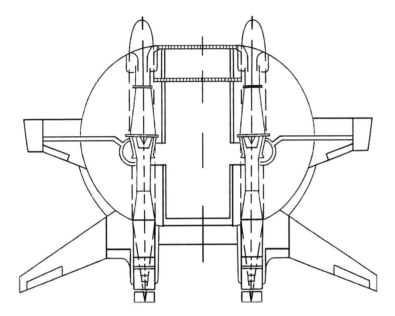

Left: **Forward and planform views of the EKIP demonstrator, showing detail of the propulsive system.** *Russian Agency for Patents & Trademarks*

Below: **An advanced passenger-carrying version of EKIP.** *EKIP Research & Production Company*

political reasons or financial difficulties. At the present time, there seems little possibility that EKIP will be developed further.

GEOBAT

The Geobat series of designs were conceived in the late 1980s by Jack Morris Jones, an American commercial artist, aviation enthusiast and experienced model aircraft designer.

His original intention was to build an airworthy remote-controlled flying saucer model, but a series of experiments with small gliders indicated that a form of 'flat annular' or 'circle' wing had a great deal in its favour, being inherently more stable at low speeds than a solid disc.

With construction of a powered prototype under way, he came up with a suitable name for the aircraft: Geobat. This is formed from the words 'geometric' and 'bat' to reflect aspects of the design's shape and appearance. The model aircraft attracted considerable attention when the first test flight took place on 14 October 1990 at the inventor's home in Georgia. Jack Jones recalls, "The whole neighbourhood was out

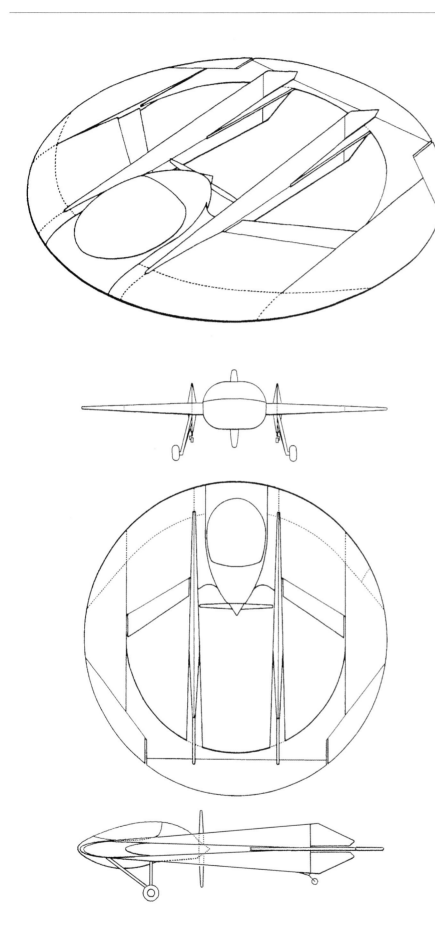

Top: **This drawing shows the Geobat in its initial form.** *US Patent Office*

Right: **An early 3-view drawing of the Geobat design. Note the proposed undercarriage layout, later changed to a tricycle configuration.** *US Patent Office*

in the streets trying to get a better view of the aircraft. They could not figure out what they were seeing, as it's difficult to make out the engine and prop."

As the new design continued to evolve, Geobat appeared to have all the makings of a successful toy and Jack Jones applied for a US patent on 16 September 1993. The story might have ended here, but the development potential for Geobat seemed obvious and the patent was revised, with a new application being filed on 12 December 1994 and published (as 5520355) during 1996.

Jones was now flying two entirely different radio-controlled Geobats. These were a 36in (914mm) diameter electrically-powered model aircraft and a larger 40in (1.016m) diameter version, driven by a small two-stroke engine. In each case, the motors were directly connected to a pusher propeller at the rear of the central fuselage section. Both Geobats were similar in appearance, featuring two separate sets of ailerons, an elevator and twin stabilising fins with rudders. After *Popular Science* magazine ran a feature on Geobat in its April 1997 issue, the project attracted plenty of interest and Jones was able to secure the attention of Senior Engineer Richard J. Foch at the US Naval Research Lab (NRL) in Washington DC.

Subsequently, Foch and two colleagues visited Jack Jones at his home near Atlanta on 26 April 1998. They discussed the design in considerable detail and observed demonstrations of the model aircraft in flight, noting Geobat's stable and predictable flying characteristics. Having expressed considerable interest in the unusual design, the NRL provided Jones with a one-year Cooperative Research and Development Agreement (CRADA).

At around the same time, Jones established Geobat Flying Saucer Aviation (GFSA) Inc with his business partner Randy Pollard, who has spent many years in the airline industry, designs model aircraft and holds a private pilot's licence. The other current member of the GFSA team is Mark Murdock, an experienced remote control model aircraft pilot, with a substantial background in civil aviation. GFSA's intention was to fully develop the design as a manned aircraft that could be built in various

Left: **Cockpit section of the 24ft (7.3m) Geobat mock-up built by Les Koberg.** *Jack M. Jones*

Below: **A model of Geobat undergoing tests in Auburn University's 3ft x 4ft Low-Subsonic Closed Circuit Wind Tunnel.** *Auburn University*

configurations to undertake a wide range of roles from a light utility aircraft to a compact transporter.

Several projects were started, including the construction of a small manned mock-up by Les Koberg, but this was considered premature and assembly was put on hold, funding being diverted to a new remote-controlled demonstrator provisionally called UAV Geobat. (UAV refers to the term 'Unmanned Aerial Vehicle'.) At the same time, a number of model aircraft enthusiasts began to draw on official Geobat plans and built their own flying versions, which ranged

from tiny micro-vehicles with the diameter of a DVD disc to substantial model aircraft. Feedback from the model aircraft community has undoubtedly been very valuable to the company. GFSA also arranged wind tunnel testing of UAV Geobat's wing section at Georgia Tech Research Institute (GTRI) that recorded good flying characteristics up to 110mph (177kph).

This was accompanied by an assessment of the design conducted by Bryan Recktenwald at Auburn University, Alabama. He carried out a series of detailed comparisons between models of the Geobat

and the Cessna 172, which confirmed lower minimum drag for the Geobat, better stall characteristics and good stability.

In early 2007, construction specialist Rick R. Hamel, based at Pittsburgh, Pennsylvania, was commissioned by GFSA to build UAV Geobat and assembly was completed one year later. The model has an 80in (1.52m) overall diameter with a flying weight of 50lb (22.6kg) and is powered by a German 5hp (3.72kW) two-stroke engine directly coupled to a two-blade 20in (508mm) pusher propeller mounted at the rear of the central fuselage section. A 6-volt electrical system was selected for motor ignition, servos and flight control, supported by nickel-cadmium batteries. Fibreglass with a Nomex core was chosen for the aircraft's skin and the wing aerofoil was based on the NACA 23012 specification. The control surfaces consist of ailerons and a centrally positioned elevator split into two units to provide a degree of redundancy in the event of technical difficulties. In addition, the UAV Geobat is equipped with two upright stabilising fins. One other feature of importance is the fully retractable tricycle undercarriage, with struts designed to absorb the impact of an exceptionally hard landing and break to protect the rest of the aircraft from serious damage.

Since trials began, take-off speed has been accurately measured at 38mph (61kph), with a cruise speed of about 50mph (80kph). The aircraft is currently limited to a maximum speed of 110mph (177kph). This may increase if a proposed change of propulsion takes place. The stall speed is calculated at approximately 25mph (40kph) with the flaps set for landing at 30°. According to Mark Murdock, who has been responsible for flight-testing, the UAV Geobat has exceeded all expectations and consistently demonstrates exceptional handling. GFSA now intends to build a manned prototype based on the same general design as UAV Geobat. This is expected to have excellent STOL performance and outstanding handling characteristics.

The first manned Geobat will have an overall diameter of 24ft (7.3m) and an airframe built from aluminium alloy, with the probable use of some composite materials. A Lycoming AEIO-360 4-cylinder engine providing 200-210hp (149-156kW) is the initial

choice for propulsion, and Jack Jones believes that an unrefuelled range of 1,000 miles (1,600km) is realistic. The prototype will cruise at approximately 170mph (273kph) and the stall speed is expected to be 45mph (72kph) or less. Maximum altitude for this aircraft will be set at 12,000ft (3,657m).

A fully retractable tricycle undercarriage will be fitted and the forward-positioned side-by-side cockpit should afford excellent visibility, although there are currently plans to improve downward visibility by fitting transparent panels. Control surfaces will be the same as found on UAV Geobat, along with the twin stabilising fins.

Equipped with a transparent Lexan leading edge lit by LEDs, the aircraft would be an impressive sight at dusk, presenting itself as a glowing arc of light in the sky. When combined with lower mounted spotlights or lasers, this has plenty of potential for use at a major event where something very unusual is required to capture the interest of spectators. According to Jack Jones, the use of Geobat

Above: **UAV Geobat during assembly in 2008.**
Jack M. Jones

Middle: **UAV Geobat during early flight trials.**
Jack M. Jones

Bottom: **An advanced future Stealth Geobat concept intended for various military applications. Engines are deeply embedded within the body of the aircraft and the vertical fins have been dispensed with.** *Jack M. Jones*

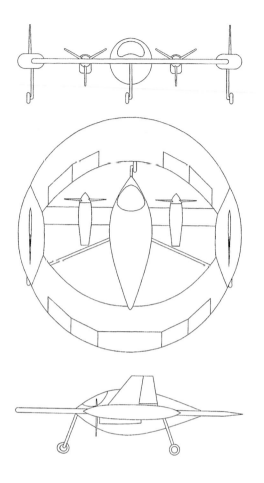

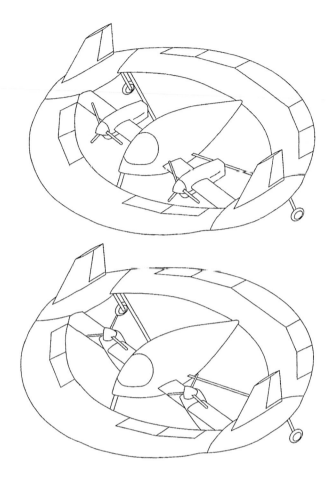

during a large open-air event has already been requested, but a suitable aircraft was not available at that time.

Even larger manned versions of Geobat have already been designed, with the possibility of a 40ft (12.1m) diameter aircraft, powered by jet propulsion. Future military applications are another possibility and Geobat is thought to be stealthy in its existing form, with considerable scope for refinement. A Micro-Geobat may be technically feasible as a surveillance tool and a larger Geobat might be developed as an advanced long-endurance radar platform, utilising the leading edge's broad arc to house the antenna.

Nevertheless, the current aim is to produce a compact, affordable utility aircraft that is fun to fly, stable and very forgiving.

RIEKEN AIRCRAFT

Having studied the 'three wing' Geobat in some detail, William Dee Rieken, a computer scientist undertaking his PhD at the Nara Institute of Science and Technology (NAIST) in Japan, began construction of a model based on the Geobat in early 2002.

The remote control aircraft was ready for its first test flight in February 2003 and Rieken's primary objective was to determine if it would make a suitable platform to carry specialised airborne sensor equipment developed at NAIST. Such a vehicle might prove invaluable for preliminary rescue and damage assessment during disasters such as earthquakes, which are fairly common occurrences in Japan.

The demonstrator was 6ft 7in (2m) in diameter, with a flight weight of 20lb (9.07kg). It was powered by an 80cc Zenoah GT80 two-stroke engine producing approximately 6hp (4.47kW) and driving a pusher propeller. Rieken's design used the same control surface layout as Geobat and it was flown using a standard radio control unit. The model was built almost entirely from composite materials and it is understood that

the original intention was to produce several versions and experiment with different control systems and sensor packages. Rieken also investigated the possibility of launching his aircraft vertically from a 'tail sitting' position, allowing it to operate from the rooftops of tall buildings.

With tests largely concluded in 2003, Rieken began to consider the possibility of evolving the design into a machine with a full VTOL and hover capability. To develop this idea, he sought the assistance of Joerg Benscheidt in Germany.

Although the original aim was to produce an unmanned, remote control, flat annular-winged aircraft that would act as a sophisticated sensor platform, Rieken's studies quickly generated a series of interesting manned aircraft designs. Consequently, these plans were patented (US: 2007215746) and tests followed involving wind tunnel trials and at least one small model.

The new concept differs considerably from Geobat and is mechanically much more complex. It uses two engines mounted on pylons between the wing and fuselage

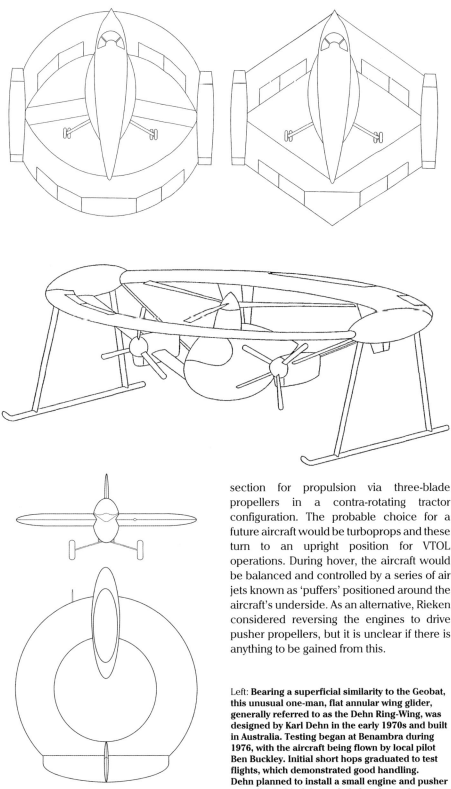

Top: **Two advanced jet-powered variants of the Rieken design. The aircraft on the left utilises a flat annular wing profile while the second drawing shows a rhomboidal option.** *US Patent Office*

Below left: **An early design for the Rieken UAV. The small aircraft shown in this illustration utilises a single three-bladed rotor for lift and two propellers for horizontal flight.** *US Patent Office*

The aircraft would also be capable of normal runway take-offs, perhaps necessary when carrying heavier payloads. To maintain balance, the cabin is positioned at the centre of the aircraft and is separated from the front wing section. Whether this would cause visibility problems from the cockpit is unclear. One reference suggests that the proposed overall diameter for Rieken's basic aircraft would be 49ft 3in (15m) and as a light transport aircraft it would have the ability to carry six passengers. All of these designs utilise ailerons and flaps along the trailing edge of the front and rear wings, an elevator to provide pitch control and stabilising fins with rudders to handle yaw. The aircraft would be fitted with a fully retractable tricycle undercarriage.

Advanced derivatives of the design include a flat annular-wing, jet-powered version intended for conventional horizontal take-off and landing. This would have a longer fuselage section that extended to the leading edge, thus improving visibility from the cockpit. This design would utilise the same control surface arrangement as the propeller-driven aircraft, although the stabilising fins with rudders would be dispensed with. Two jet engines would be mounted within wingtip pods and the aircraft would be fitted with a fully retractable tricycle undercarriage.

An alternative to this proposal would be a rhomboidal wing variant. The most likely near-term development of Rieken's design will be a UAV, utilising two contra-rotating rotors for vertical lift, plus two fixed position engines for horizontal flight. A fixed support frame will be employed for the vehicle's undercarriage, with stabilising fins with rudders positioned downwards below the wing. Equipped with an advanced 360° vision sensor system, the UAV has considerable potential for further development and at the time of writing it is understood that a company in California has been licensed to develop and build this design.

section for propulsion via three-blade propellers in a contra-rotating tractor configuration. The probable choice for a future aircraft would be turboprops and these turn to an upright position for VTOL operations. During hover, the aircraft would be balanced and controlled by a series of air jets known as 'puffers' positioned around the aircraft's underside. As an alternative, Rieken considered reversing the engines to drive pusher propellers, but it is unclear if there is anything to be gained from this.

Left: **Bearing a superficial similarity to the Geobat, this unusual one-man, flat annular wing glider, generally referred to as the Dehn Ring-Wing, was designed by Karl Dehn in the early 1970s and built in Australia. Testing began at Benambra during 1976, with the aircraft being flown by local pilot Ben Buckley. Initial short hops graduated to test flights, which demonstrated good handling. Dehn planned to install a small engine and pusher propeller behind the cockpit, but the project ended in 1982 when he died. The aircraft was then donated to the Air World Aviation Museum at Wangaratta. It is known that Dehn made various minor modifications to the design and the undercarriage was altered on a number of occasions, using different tail wheel arrangements. Specifications are not available, although the span of the glider is thought to be 16ft 5in (5m).** *Bill Rose*

A Copy of an originally classified document issued by the 970th Counter Intelligence Corps on 10 November 1947 to senior officers based in Europe. This reveals a belief that flying saucers were genuine. It says that models were being built to test their aerodynamic properties and raises the possibility that flying disc technology originated in wartime Germany.

SECRET

HEADQUARTERS
COUNTER INTELLIGENCE CORPS REGION VI
970th COUNTER INTELLIGENCE CORPS DETACHMENT
APO 159
10 November 1947
VI-13-1006
D-198239
VI 1611.5

SUBJECT: Flying Saucers

To: Commanding Officer Bayreuth Sub Region
Nurnberg Sub Region
Wurzberg Sub Region
Bamberg Sub Region
Regional Technical Specialist

1. Considerable material has been gathered by the Air Materiel Command, WRIGHT FIELD, Ohio, concerning the appearance, description and functioning of the object popularly known as "Flying Saucers". A copy of the report from the Air Materiel Command is on file at this Headquarters.

2. The opinion was expressed that some sort of object, such as the flying saucer, did exist. At the present time, construction models are being built for wind tunnel tests. It is further suspected that the flying objects may have been developed from original plans and experiments conducted by the Germans prior to the capitulation.

Headquarters, 970th CIO Detachment, European Command is desirous of locating German Aircraft specialists and test pilots who might have some knowledge of similar aircraft. It is requested you canvass your area for possible identity of aircraft specialists or test pilots known in your area with such knowledge. This canvass is to be made discreetly and to conceal our interest in the subject. If any are located, a copy of the Air Materiel Command report with specific EEI's will be forwarded to permit further interrogation and test of knowledge.

3. This canvass, both of files and sources of information, should be begun as soon as possible. Your initial report is due this Headquarters by:

Suspense Date
12 Dec 47

BY ORDER OF LT COLONEL WALKER

[Signed]

WILLIAM E LAHNED JR.
Special Agent CIC
Operations

SECRET

Horten Parabola

Accommodation: 1
Wingspan: 39ft (11.9m)
Empty weight: 198lb (90kg)
Weight max: 374lb (170kg)
Powerplant: none
Maximum speed: 101mph (163kph)
Landing speed: 18mph (29kph)
Stall speed: 18mph (29kph)
Built at Aegidienberg near Koblenz in 1938, this very light sailplane was never flown, having been badly damaged during storage. It was not considered repairable and the Parabola was set on fire and destroyed.

Loedding's VTOL Study

Accommodation: 4
Span: N/A
Length: N/A
Gross weight: N/A
Powerplant: 1 x unspecified single water-cooled internal combustion engine driving a fan, or a gas turbine
Maximum speed: Subsonic

Avro-Canada Project Y 'Ace'

Accommodation: 1 (two-seat variant considered)
Wingspan: 30ft (9.14m) approx. Later 25ft (7.62m) approx
Length: 40ft (12.19m) approx. Later 35ft (10.6m) approx
Weights: N/A
Powerplant: 1 x first generation Orenda radial flow gas turbine. Thrust unknown
Maximum speed: Mach 2.25 in level flight
Ceiling: 65,000ft (19,812m)
Range: 1,000 miles (1,609km) in clean condition
Armament: Options considered included two or four cannons and/or a pack containing spin stabilised Folding Fin Aircraft Rockets (FFAR). A future possibility was two de Havilland Blue Jay air-to-air missiles on external rails.

Avro 724 Version A

Accommodation: 1
Wingspan: 16ft (4.88m)
Length: 32ft 6in (9.91m)
Powerplant: 1 x Rolls-Royce RB.106 axial flow turbojet engine, providing 15,000lb (66kN) static thrust and 21,800lb (97kN) using the afterburner
Maximum anticipated speed: Mach 2
Armament: not specified, probably 2 x air-to-air missiles

Avro 724 Version B

Accommodation: 1
Span: 24ft (7.32m)
Sweep: 60° leading edge
Length: 37ft (11.28m)
Gross weight: 24,500lb (11,113kg)
Powerplant: 2 x Rolls-Royce RB.106 axial flow turbojet engines, each providing 15,000lb (66kN) static thrust and 21,800lb (97kN) using the afterburner
Maximum speed: Mach 2.5
Ceiling: 60,000ft+ (18,288m+)
Armament: 2 x air-to-air missiles

Avro-Canada Project Y2 RFGT Aircraft

Accommodation: 1
Diameter: 29ft (8.84m)
Powerplant: Improved Orenda radial flow gas turbine. Thrust unspecified
Gross weight: 29,000lb (13,154kg)
Maximum speed: Mach 3.5
Ceiling: 80,000ft+ (24,384m+)
Range: 620 miles (997km)
Armament: 4 x 20mm cannons. Possible armoured edge for ramming enemy aircraft

Avro-Canada (MX) 1794

Accommodation: 1
Diameter: 35ft 3.5in (10.76m)
Height: 7ft 8.5in (2.35m)
Powerplant: Turbo-ramjet using 6 radially dispersed Viper turbojets
Maximum speed: Mach 3.5-4.0
Rate of climb: 3 minutes to 75,000ft (22,860m) from hover
Ceiling: 105,000ft (32,000m)
Range: 1,000 miles (1,609km)
Armament: 20mm rotary cannon. 2 x air-to-air missiles

Avro-Canada WS-606A Configuration A

Crew: 1
Wingspan: 29ft (8.84m)
Wing area: 1,083sq ft (100.6m^2)
Length: 37ft (11.28m)
Height: 7ft 8in (2.33m)
Gross weight: 65,000lb (29,483.5kg)
Powerplant: 6 x Viper turbojets

Avro-Canada WS-606A Configuration B

Crew: 2
Similar gross weight to Configuration A
Powerplant: 2 x (P&W or General Electric) afterburning turbojets, producing 55,000lb (244kN) thrust
Maximum speed: Mach 2
Combat radius: 500 miles (804km) using VTOL; 700 miles (1,126km) with GETO
Armament (interceptor): 2 x air-to-air missiles

Avrocar

Accommodation: 2
Diameter: 18ft (5.49m)
Height: 3ft (0.91m)
Weight: 5,650lb (2,562kg)
Propulsion: 3 x symmetrically disposed Continental J-69 turbojets, each rated at 927lb (4.12kN) static thrust
Central fan diameter: 5ft (1.52m)
Maximum speed (predicted): 300mph (482kph)
Ceiling (predicted): 15,000ft
Range (predicted): 1,000 miles (1,609km)
Payload (predicted): 2,000lb (907kg)

Lockheed Supersonic Disc

Crew (military role): 2 (perhaps up to 4)
Diameter: 50ft (15.24m)
Gross weight: 55,000lb (24,947kg)
Powerplant: Turbo-ramjet system
Maximum speed: Mach 4
Ceiling: 100,000ft (30,480m)
Range: 7,500 miles (12,070km) using extra reserves of hydrogen fuel
Armament: 2 x free-fall nuclear weapons?

Couzinet RC.360

Accommodation: 1 or 2
Diameter: 44ft 7in (13.6m)
Lifting surface area: 645.6sq ft (60.0m2)
Gross weight: 27,700lb (12,565kg)
Propulsion: 6 x Lycoming engines driving two 50-blade rotors. For level flight, 1 x Armstrong-Siddeley Viper turbojet, delivering 1,639lb (7.2kN) of static thrust
Maximum speed: N/A
Ceiling: N/A
Range: N/A
Armament: N/A

Ling-Temco-Vought V-468 ADAM II (1965)

Accommodation: 1
Span: 31ft 8in (9.65m)
Wing area: 291sq ft (27m²)
Length: 45ft (13.7m)
Height: 10ft 6in (3.2m)
Empty weight: 12,300lb (5,579kg)
Gross weight: 17,400lb (7,892.5kg)
Overload weight: 20,300lb (9,208kg)
Powerplant: 2 x GE1/J1 Gas Generators shaft-coupled to four wing fans and small nose fan
Maximum speed: Mach 0.89
Ceiling: N/A
Range: N/A
Armament: Wide range of external stores and probably air-to-air missiles for self-defence

Tumenecotrans Bella-1

Accommodation: 2 plus 4 passengers
Wingspan: 36ft 1in (11m)
Central disc diameter: 15ft (4.6m)
Length: 32.8ft (10m)
Height: 8ft 6in (2.6m)
Gross weight: 4,400lb (2,000kg)
Powerplant: For horizontal flight, 2 x externally mounted Teledyne Continental IO-360 6-cylinder, fuel injected engines each rated at 210hp (156.5kW). For lift, 1 x Teledyne Continental 4-cylinder air-cooled horizontally opposed piston engines rated at 100hp (74.5kW)
Cruise speed: 155mph (250kph)
Ceiling: 10,000ft (3,000m)
Range: 620 miles (1,000km) with maximum payload
Payload capability: 1,323lb (600kg)

UAV Geobat

Wingspan: 80in (1.52m)
Wing area: 17.5sq ft (1.62m²)
Length: 80in (1.52m)
Height: 13in (330mm)
Gross weight: 50lb (22.6kg)
Powerplant: 5hp (3.72kW) two-stroke engine driving a two-blade 20in (508mm) pusher propeller
Cruise speed: 50mph (80kph)
Maximum speed: 110mph (177kph)
Ceiling: 12,000ft (3,657m) with current propulsion system
Payload capability: N/A

Accidental Sightings and Secret Tests

Were any post-war disc planes built and test-flown? This question has continually recurred since I first became interested in flying saucers and like the majority of serious aviation and space writers, my initial reaction has been to dismiss the idea as unlikely and unproven.

However, if we briefly step into the world of conspiracy theory, it is a fact that many major US defence contractors studied flying saucer aircraft during the immediate post-war years. The list includes Boeing, Douglas, Convair, Republic North American and Lockheed. But, in many cases, the details of these projects have never been fully disclosed, perhaps because any association with UFOs would be regarded as undesirable. Nevertheless, there is a suspicion that some ideas progressed beyond the drawing board and a small number of man-made flying saucers could have been built and tested in remote parts of the United States.

One reasonable argument sometimes made against the testing of disc-shaped prototypes is their absence in modern aviation. If they flew, why are there none in use today? The answer might be that the flying saucer was a concept that caught the

imagination of aeronautical designers but proved too far ahead of its time to be workable. It could also be argued that the history of man-made flying saucers has been kept secret to retain the useful cover provided by UFOs. Let me just add that I do not have any inside information to make a case for this or any other possibility and my objective is just to present ideas for readers to consider.

THE RHODES PHOTOGRAPHS

One particularly intriguing UFO report from Phoenix, Arizona, makes a strong case for the existence of a flying saucer shaped prototype. It was late in the afternoon of 7 July 1947 when the sound of a passing jet aircraft caused William Albert Rhodes (1916-2007) to look up from his backyard. He expected to see a fighter like the Lockheed P-80 Shooting Star, but the aircraft crossing the sky had a very unusual heel-shaped appearance. Rhodes estimated the aircraft's diameter to be 20-30ft (6-9m). Its initial altitude was about 5,000ft (1,500m) and he thought it was travelling at 400-600mph (640-960kph). It should be noted that the accuracy of detail provided by most untrained observers is often

rather questionable.

However, Rhodes' description of the unidentified aircraft was reasonably detailed. It was grey and had a clearly visible cockpit canopy in the centre. No propellers were fitted and the initial jet engine noise disappeared. The aircraft then descended to about 1,000-2,000ft (300-600m) and circled in the east for approximately 30 seconds. While this was taking place, he ran to get a Kodak Box Brownie (620 format) camera from his workshop and secured two photographs before his film ran out and the craft was lost from sight.

That evening, Rhodes processed the black and white film and made some prints of the unidentified aircraft and the following day he contacted *The Arizona Republic* newspaper with his story. The newspaper published the details on 9 July 1947, along with blow-ups of the alleged aircraft that appeared fairly distinct and genuine. The authorities were now aware of this sighting and on 29 August 1947, Rhodes received a visit from an Army counter-intelligence (CIC) officer and an FBI agent who asked him to provide the two negatives for his photographs. He complied the following day, taking his negatives to the FBI office in Phoenix – but that would be the last time he ever saw them.

Rhodes was investigated and background checks were made, while scientific analysis determined that the object was genuine and had a diagonal length of about 40-50ft (12-15m). He was officially interviewed at his home for a second time on 11 May 1948 by Project Sign representatives Alfred Loedding and Lieutenant Colonel James C. Beam. Despite attempts to secure possession of his

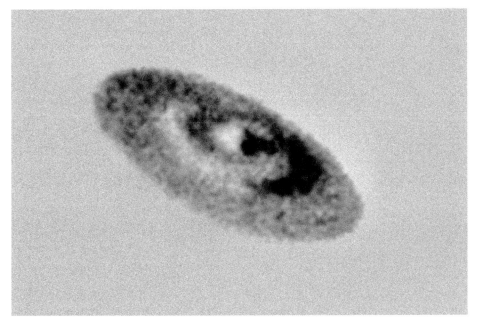

Left: **The object seen in this sectional enlargement remains unidentified. It was captured in five photographs taken by press photographer Ed Keffel at Barra da Tijuca near Rio de Janeiro, Brazil, on 7 May 1952. The medium format images have undergone considerable analysis and are generally regarded as authentic. It would be difficult to produce similar fakes, but technical issues with the lighting have cast doubt on this material. It is interesting to note the similar appearance of the object to a saucer seen in the 1954 SF movie *This Island Earth* and designs by Francesco de Beaumont and John Frost.** *Ed Keffel*

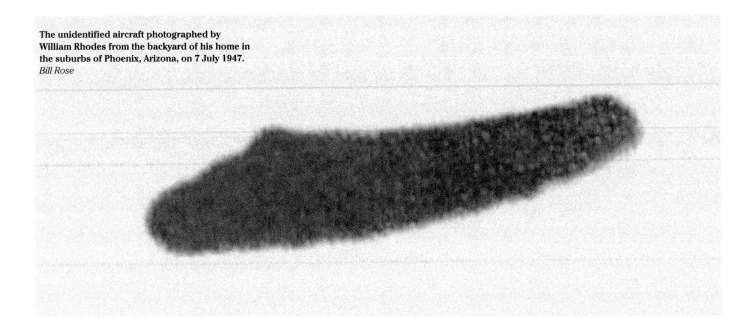

The unidentified aircraft photographed by William Rhodes from the backyard of his home in the suburbs of Phoenix, Arizona, on 7 July 1947. *Bill Rose*

negatives, they were never returned and Rhodes refused to participate in further interviews. Loedding had been largely convinced by the authenticity of the photographs and, interestingly, Arnold was shown the pictures and remarked that the craft looked like one of the discs from his own sighting.

Rhodes was a musician with an interest in science. He was an amateur experimenter who claimed to have a PhD in Physics, which was never verified. His activities were sufficiently unusual to provide the air force with scope to eventually dismiss the sighting as a probable hoax.

If this was the chance sighting of a secret prototype aircraft, it is interesting to compare its appearance with later designs for the Project Y interceptor. Assuming that this sighting was genuine, it may indicate an attempt to secretly develop and exploit very unusual technology captured from the

Germans, such as an RFGT engine. Other reports from the same area appear to confirm this was a genuine sighting, but there is still insufficient evidence to be absolutely certain.

AN UNEXPLAINED EVENT IN EAST GERMANY

If there is any truth to the information filed by Andreas Epp that a flying saucer was developed in post-war East Germany, an accidental sighting may provide some support. This incident is widely reported to have taken place on 9 July 1952 (the same day that an account of the Spitsbergen saucer appeared in *Berliner Volksblatt*), but the principal witness later changed the date to 17 June 1950, blaming the publishers for the error.

The details of this unusual incident were provided by Oscar Linke (1904-?), a Wehrmacht Major during the war and the

Mayor of Gleimershausen. Having escaped to West Berlin with his family, Linke reported the event to local officials and was interviewed at some length by Western intelligence officers. According to Linke, he had been travelling home on his motorcycle accompanied by his 11-year-old daughter Gabriella. A punctured tyre forced them to complete the remainder of the journey to Hasselbach on foot. This was rather unwelcome with twilight approaching.

Walking near a heavily wooded area, Gabriella spotted what she thought to be a deer beyond some trees. Cautiously approaching to gain a better view, it became clear that they were actually looking at two men in a clearing, wearing metallic-looking suits. The men were apparently studying something on the ground and one had some kind of a light attached to his chest.

From their concealed vantage point some 150ft (50m) from the men it was also possible to see a substantial pan-shaped object standing on the ground behind them, which Linke estimated to have a diameter of about 40-50ft (12-15m). There were precisely spaced holes around the rim of this device that had a supporting cylindrical section at its centre with an estimated height of about 10ft (3m).

According to Linke, his daughter called out in surprise and this caught the men's attention. They immediately went to the

Left: **This image allegedly snatched by William Rhodes in 1947 provides a clearer impression of the unidentified object's shape. Interestingly, it bears quite a resemblance to the later Avro-Canada Project Y design.** *Bill Rose*

Flying Saucer Technology

mysterious object, climbed through a hatch and moments later the object's rim began to glow and then rotate. The central cylindrical section retracted and the machine lifted off from the ground. An initial humming noise gave way to a roar, with visible flames around the rim. The craft briefly hovered and then began to fly horizontally in the direction of Stockheim to the south-west of their location. The machine continued to gain altitude until it was lost from sight, but apparently there were several sightings of the object by local people.

Believing it to be a new experimental Russian aircraft, Linke ensured that his daughter kept the story to herself for the remainder of their time in East Germany. According to Linke, people could be imprisoned for years if they knew too much about the wrong things.

While the object described differs considerably from the designs that Epp claimed to have worked on, this incident remains interesting. If it was genuine, the sighting could indicate the testing of a vehicle derived from wartime Nazi research. If the story

is false, it may have been contrived by Linke in an attempt to win concessions from the West German authorities. Alternatively, he could have been used by a Western intelligence agency to influence public opinion and help ensure increased defence spending.

However, the general consensus of opinion seems to suggest that Linke was telling the truth and the incident was real, making this one of the more interesting UFO reports that appears to involve secret man-made machinery.

THE RUSSELL SIGHTING
On 4 October 1955, four members of a high-level American delegation were travelling on a train through the Soviet Trans-Caucasus region. Heading this group was Republican Senator Richard Brevard Russell Jr (1897-1971) who led the Senate Armed Forces Committee. One of America's foremost experts on military and defence policy, Russell had unsuccessfully sought the Democratic Presidential nomination in 1952. Accompanying the senator was his military

aide, US Army Lieutenant Colonel E. U. Hathaway, Reuben Efron, a Russian interpreter who also kept records of their activities, and a fourth unidentified person, who may have been a CIA officer.

Much of the men's time was spent idly observing the passing countryside on their journey between Atjaty and Adzhijabul. It was just after 7.00pm and twilight had arrived, although visibility from the carriage was apparently excellent. Russell was alone in his compartment staring out of the window when he suddenly became aware of a saucer-shaped aircraft climbing into the sky from an obscured area about 1 mile (1.6km) distant.

Russell now alerted his assistants in the next compartment. They observed the unidentified aircraft moving across the sky and about one minute later it was joined by an identical vehicle that ascended from the same take-off area. Each disc was metallic in colour and revolved clockwise, although the lights on top of the craft seemed stationary.

There appeared to be some flame from beneath each disc and there were visible exhaust emissions.

A simulated image of an unidentified flying object produced by the author. *Bill Rose*

The mysterious craft moved towards the train, and the American team were certain that searchlights on the ground were being used to illuminate them. The discs now picked up speed and soon passed over the train, heading north at an estimated altitude of 6,000ft (1,829m). Apparently, at this point a railway official entered the compartment and insisted on lowering the window blinds, which prevented any further observations.

Several days later, the delegation arrived at the US Embassy in Prague, Czech-oslovakia, and reported the event to the Air Attaché Lieutenant Colonel Thomas Ryan. He immediately forwarded a top-secret summary of the incident to Washington and this was passed to the CIA. On 19 October 1955, the CIA's Chief of Scientific Intelligence, W. E. Lexow, filed a classified memorandum that compared the aircraft seen from the train to Avro-Canada's secret Project Y design.

A high-level delegation of American officials travelling through Soviet territory at the height of the Cold War would have been closely monitored and controlled by the KGB (now the FSB) who handled all aspects of counter-intelligence, internal security and surveillance. Even today, any important foreign delegation would receive special attention from the Security Service, but at that time it seems inconceivable that they would have been allowed to see anything unusual unless it was intentional. The account of this incident has to be treated as totally reliable and there seems no reason to dispute the details.

Was the event carefully staged by the Russians to make a specific impression on the Russell delegation? If so, it certainly sent a strong signal to Washington that the Russians could match any new developments taking place in the West and might even be some way ahead. Assuming these were Russian-built flying saucer aircraft, it indicates there probably was a project stemming from wartime German research that had reached the flight-testing stage. If this was the case, we can only assume that the technology proved troublesome or impractical and was eventually abandoned. The Soviets were well aware of the West's obsession with flying saucers and this could have been part of a

Above: **US Senator Richard Brevard Russell Jr, who observed two flying saucers in Soviet territory during 1955.** *US Congress*

Below: **A re-creation of how the two unidentified objects seen by the Russell delegation may have appeared.** *Bill Rose*

Flying Saucer Technology

complex psychological warfare programme undertaken over many years.

Whether the truth about any of this will surface in Russia at some point in the future is unknown. As for the American details of the sighting, these remained classified until 1985.

AVRO-CANADA SPECULATION

Although flying saucer shaped aircraft were designed and evaluated as wind tunnel models by most of the major US defence contractors, it was Avro-Canada who reached the hardware stage. There are still gaps in the history of Project Y2, MX 1794 and WS-606A. It is rumoured that mock-ups of all

three were built and at least two prototypes were largely completed. But there are no photographs or details of this work, which was apparently abandoned.

Some conspiracy theorists believe that final development of the Avro-Canada supersonic saucers may have been transferred to a major US contractor. Documents can be contradictory, but it is known that the USAF decided to proceed with development of a Project Y2 aircraft on 29 December 1954 and the British anticipated completion of the Y2 research vehicle in 1955. Project Y2 rapidly evolved into Project MX 1794 during 1955 and by the following year, a turbo-ramjet engine test rig had been completed.

USAF document 57WCLS-2497 reveals that on 26 March 1957 a delegation of senior USAF officials visited Avro-Canada and were shown 'facilities being used to manufacture the (1794) prototype aircraft'. The report mentions that 'this (tour) included an inspection of various (1794) components which have already been produced'. Another previously classified USAF document from 1957 (reference 57RDZ-3107) tells us that 'the design of the basic structure and propulsion unit (of 1794) is substantially complete and it is expected that the additional items together with all preliminary modifications to the rig design required by the prototype will be complete by June 1957'. This document goes on to mention the use of a tethered test stage and tethering rig for preliminary flight trials, 'probably consisting mainly of a large crane'. It is accepted that there were serious problems with the engine tests and nobody would deny that Frost's designs were extremely complex and contained features that were unreliable and prone to sudden failure.

But the possibility of research and development continuing at a secluded American location like Groom Dry Lake in Nevada cannot be entirely ruled out, and Lockheed's Skunk Works who participated in the development of this USAF/CIA facility would have been the perfect organisation to continue working on a black project of this nature.

Aside from the USAF, CIA, specialist contractors and senior US politicians, nobody has any real idea what takes place at Area 51 Groom Dry Lake. The long-term history of this multi-billion dollar research facility is patchy at best and for many years the Pentagon denied that it even existed. Many UFO sightings have been reported around the Groom Dry Lake area, leading to claims that alien spacecraft have been tested at this base for more than half a century.

Whether any UFO sightings could be linked to John Frost's project during the mid-1950s is impossible to say, but it cannot be entirely ruled out.

Top: **This photo-simulation shows an Avro-Canada MX 1794 undergoing secret trials at a remote USAF test facility such as Area 51.** *Bill Rose*

Left: **This photomontage shows the probable forward appearance of an Avro-Canada MX 1794 prototype during testing at a secret USAF site.** *Bill Rose*

Chapter Five

Lighter Than Air

This chapter briefly explores the links between balloons, airships and UFOs. It also examines a number of lighter-than-air craft with the appearance of flying saucers. From late Victorian times, there were many sightings of mysterious unidentified flying objects in the skies around the world, with the majority occurring in America. Manned flight remained a novelty and there was a consensus of opinion that a small number of secretly built airships were responsible.

Many sightings were obvious hoaxes created by the media to improve newspaper circulation figures, while others arose from simple misidentification or individuals seeking publicity. On the other hand, it is probable that some of these reports can be attributed to test flights of privately built experimental airships that have never been revealed. Naturally, those ufologists with a leaning towards alien visitation often have very different interpretations of these pre-Arnold sightings.

There were several 'waves' of unexplained airship sightings across the United States during 1896 and 1897, and this also occurred to a lesser extent in some other countries. Although the inventor Dr Solomon Andrews (1806-1872) had flown the first American-built, hydrogen-filled, steerable airship above Perth Amboy, New Jersey, on 1 June 1863, the sight of balloons and specifically dirigible airships in American skies remained unusual.

The 1896 airship sightings began in California and spread eastward across the country, with many accounts describing glimpses of occupants in the craft and the use of powerful searchlights after dark. Since then, a handful of inventors and aviation pioneers have been identified as the possible perpetrators of these mysterious flights. Their names include Lyman Gilmore (1874-1951) and Charles Dellschau (1830-1923) who apparently belonged to a secret society called

Below: **The world's first steam-powered airship built by Henri Giffard took to the skies on 24 September 1852, flying from Paris to Trappes. Although Giffard was unable to return to the take-off location, he demonstrated the ability to manoeuvre the craft.** *Bill Rose*

the Sonora Aero Club. Although no hard evidence exists to show that airships were secretly built by these individuals, the possibility cannot be entirely ruled out.

Conceivably, some reports of mysterious airships may have been triggered by the 1886 Jules Verne SF novel *Robur the Conqueror*, which fired the American public's imagination with the tale of a renegade inventor who built and operated a long-range airship. Similar public responses have arisen on several occasions since then, created by films such as *The Day the Earth Stood Still*, *Close Encounters of the Third Kind* and the TV series *The X-Files* (1993-2002). All produced a huge increase in UFO sightings and – during *The X-Files* period – claims of alien abduction.

The 19th century American airship sightings peaked during 1897, with a crash landing that has much in common with the infamous Roswell Incident which occurred half a century later. Allegedly, an unidentified airship collided with a windmill at Aurora, Texas, on 17 April 1897. The craft is said to have been totally destroyed and the pilot killed. According to a report in the *Dallas Morning News*, the pilot was 'not an inhabitant of this world' and Mr. T. J. Weems, a US Army officer who attended the scene, believed he was probably a native of the planet Mars.

This was an interesting remark, as the SF novel *War of the Worlds* by H. G. Wells, which describes a Martian invasion, was not published until the following year. It is also hard to understand why any race of spacefaring aliens would use crude Victorian technology to travel around our planet.

Apparently, the alien was buried with Christian rites in the local cemetery and the airship wreckage was dumped into a disused well and covered over. Unlike Roswell, there was no government involvement or official cover-up and the story is generally accepted to have been a hoax, perpetrated by some members of the local community who were attempting to boost the failing local economy.

Sightings of mysterious airships continued to be reported across the United States, but public interest began to wane and the media moved on. However, a steady number of airship sightings continued to be made in other parts of the world, including Britain.

This trickle of unexplained reports persisted throughout the first decade of the next century until a British police constable on patrol in Peterborough during the early hours of 23 March 1909 reported seeing an unknown powered airship with searchlights.

The British press seized on this story and it

led to what would be called the 'Airship Scare', sightings being made across the country and in particular East Anglia. There was now increasing tension with Germany that would lead to World War 1 and there were regular suggestions in the press that these craft were being used to spy on Britain. The reports continued for several years, but no evidence has ever emerged to prove the existence of these airships.

By the late 1940s, there was something close to a re-run of these events taking place, but this time the airships had been replaced by alien spaceships. Now there was a global willingness to accept the possibility that extraterrestrials were visiting our planet, while Western military leaders secretly feared that advanced Nazi flying disc technology had fallen into Soviet hands.

Above: **The first flight of Zeppelin LZ1 at Lake Constance (Bodensee) in July 1900. This was more than four years after the first reports of unidentified airships above regions of the United States. Early 20th century reports of mystery airships above Britain were often believed to be German Zeppelins on spy flights, but no proof of this has ever surfaced.** *Bill Rose*

Below: **The details of this Italian saucer-shaped airship are unknown and the illustration may show a fictional concept or a model. It is claimed that the airship was secretly built during the Mussolini era, but no details were available at the time of writing this book.** *Unknown. Restoration by Bill Rose*

ROSWELL

When Kenneth Arnold reported his sighting of unidentified objects crossing the Cascade Mountains in June 1947, the story was picked up by the media and amplified into something new that would capture the public's imagination. Just days later, a second event would overshadow this and eventually become a modern-day legend.

On 3 July 1947, William 'Mac' Brazel (1899-1963) who managed the J. B. Foster Ranch near Corona was checking his fields on horseback when he discovered a substantial quantity of unusual debris scattered across one of his fields. Brazel believed he had found the remains of an aircraft that had run into trouble during a severe thunderstorm that had swept across the area during the previous night. The exact details of the story differ slightly depending on the source and there is some disagreement on the dates. But it seems that Brazel reported his discovery to the Sheriff's Office in nearby Roswell on 5 July, and in turn, Sheriff George Wilcox contacted the USAAF at the Roswell Army Air Field, which was home to the 509th Bomb Group.

As a result, Intelligence Officer Major Jesse Marcel and Counter-intelligence Corps official Sheridan Cavitt were sent to the ranch on 7 July. Brazel showed them the debris field and officially they spent several hours gathering up pieces of wreckage, which consisted of tin foil, rubber strips, some paper and short pieces of wood. Apparently, this amounted to a bundle of material weighing about 5lb (2.26kg), which Marcel took back to Roswell AAF.

That should have been the end of the matter, but early on 8 July 1947, Colonel William H. Blanchard, who commanded the 509th Bomb Group at Roswell, approved a press release stating that a flying saucer had been recovered by the USAAF. It read: 'The many rumours regarding the flying disc became a reality yesterday when the intelligence office of the 509th Bomb group of the Eighth Air Force, Roswell Army Air Field, was fortunate enough to gain possession of a disc through the cooperation of one of the local ranchers and the sheriff's office of Chaves County.'

The media immediately picked up on these details and one of the first newspapers to break the story was the *Roswell Daily Record* which announced on its front page, 'RAAF Captures Flying Saucer on Ranch in Roswell Region'. Why this astonishingly ill-considered press release was issued remains unclear, but the effect was predictable and within hours the USAAF had a major public relations problem on their hands.

Senior USAAF officials' hurriedly arranged a press conference at Fort Worth where they displayed debris described as pieces of a weather balloon recovered from a Roswell ranch. Eventually, the public lost interest in this story. Many years later, the incident underwent a major revival with claims that a large debris field had been discovered, along with part of a flying saucer and several alien bodies. Allegations were now being made that a huge cover-up had taken place and these were endorsed by an ageing Marcel and various other witnesses. Clearly something unusual had occurred at Roswell in 1947 and although many of the new accounts seemed implausible, the story simply would not go away.

During the early 1990s, Republican Congressman Steven H. Schiff (1947-1998) became interested in the Roswell Incident and he requested all official documents relating to it. Schiff met with considerable resistance, and in 1994, he used the Congressional General Accounting Office to force the USAF to reopen the case. This resulted in the USAF commissioning an investigation team headed by Colonel Richard Weaver who eventually admitted that there had been a cover-up and suggested that the recovery of top-secret balloon equipment had been the cause.

The balloon programme was called Project Mogul and Weaver's team believed

Left: **On the afternoon of 8 July 1947, a press conference was hastily convened in Brigadier General Roger M. Ramey's Office at Fort Worth Army Air Field. The purpose of this gathering was to explain the situation at Roswell following reports that debris from a crashed flying saucer had been secured. The USAAF now decided to announce that a weather balloon was responsible for the incident and pieces of this had been recovered from the Foster Ranch. In this photograph taken by the late James B. Johnson, USAAF Intelligence Officer Major Jesse Marcel displays pieces of a weather balloon. However, when the case was reopened fifty years later, the USAF acknowledged this had been a cover-up and suggested this was done to conceal the recovery of a top-secret balloon called Mogul. Many investigators remain unconvinced that this is the true explanation.** *Fort Worth Star-Telegram Photograph Collection*, *Special Collections Division*, *University of Texas at Arlington Libraries*.

that a service flight (Number Four launched on 4 June 1947) from Alamogordo, New Mexico, was probably responsible for the incident. The purpose of Project Mogul was long-range acoustic detection of Soviet nuclear explosions at high altitudes. It was theorised that this particular Mogul balloon train had been caught by unusual winds and carried off-course to the Foster Ranch, where Brazel discovered the remains. However, this suggested explanation was seen by many as another whitewash and numerous key documents were said to have been accidentally destroyed after the incident, raising further doubts about official honesty.

The initial Air Force re-examination of the Roswell Incident met with ridicule within the UFO community and this prompted a second release on 24 June 1997, titled 'The Roswell Report: Case Closed'. This reaffirmed the Project Mogul balloon theory, although it is said that the USAF considered a second explanation involving the crash of a highly classified glider/balloon vehicle. This report simply convinced UFO investigators that the authorities were continuing to cover up something of major importance, and the belief that a flying saucer complete with alien bodies was recovered was now firmly embedded in many people's minds. What exactly took place on the Foster Ranch in July 1947 remains a mystery to everyone outside the highest levels of US government.

The Mogul balloon theory appears to be flawed, but the USAF tried hard to dismiss all other possibilities. These include the crash landing of a highly classified aircraft, rocket or an early Broken Arrow nuclear weapon accident. And, of course, it almost goes without saying, the discovery of a crashed alien spacecraft. The Roswell Incident remains intriguing. The USAF has failed to explain why the recovery of a balloon train carrying primitive acoustic sensor equipment would need to be hidden behind an impenetrable wall of secrecy for half a century.

That said, it is not my intention to delve too deeply into this case as there are numerous publications available on the subject and I do not want to stray from the general theme of this book. My aim at the start of this chapter is to establish the link between UFO/flying saucer reports and lighter-than-air craft. What seems evident is that, since 1947, several US government agencies have used interest in UFOs as a way of discrediting accidental sightings of highly classified balloons and, later, specialised aircraft.

During the early 1950s, the CIA identified the potential threat posed by public interest in UFOs, recommending that reports should be discredited and UFO groups placed under surveillance. These recommendations filtered down through all US military departments and were followed by new regulations prohibiting public disclosure of UFO sightings. At the same time, it became easier to hide the accidental observation of a secret operation behind a readily available means to ridicule observers should this become necessary. Balloon technology used by the US military was simple and hardly secret, but the payloads were becoming increasingly sophisticated and highly sensitive.

While it could be argued that balloons have no place in a book that sets out to deal with flying saucer technology, I feel it will fill in some useful gaps to devote a few pages in this final chapter to the Mogul, Skyhook and Genetrix balloon projects. These have links to major UFO incidents and were responsible for countless reports of flying saucers across North America, Europe and Russia throughout much of the Cold War era.

PROJECT MOGUL

This largely unknown project was officially identified in the 1990s as being responsible for the Roswell Incident. Mogul had originated in the immediate post-war years when US Intelligence realised that the Soviets were undertaking a crash programme to develop nuclear weapons.

The Russians were not expected to produce their first atomic device before 1950, but the Central Intelligence Group (the initial name for the CIA) was becoming increasingly concerned about long-range detection of a nuclear explosion within Soviet territory.

Gathering information on military developments behind the Iron Curtain was extremely difficult at this time, but the issue of detecting a Russian atomic test was given priority and a committee was formed to consider the issue.

One interesting proposal came from by Dr Maurice Ewing (1906-1974) of Columbia University, NY, who was a leading expert on acoustic and seismic events. He had studied ways of detecting distant underwater explosions via sound channels during World War 2 and believed that something similar might take place in the lower stratosphere, where air pressure and temperature allowed

Right: **Project Mogul Balloon Train for Cluster Flight No 2.** *USAF*

the ducting of low frequency sound waves over long distances.

Ewing suggested that the signature of a nuclear explosion might be detected by suitable equipment carried by a balloon that could maintain a constant height. As a consequence, the committee decided to investigate the idea further. The New York University (NYU) was undertaking a balloon programme and were consulted about developing a suitable system to lift an unspecified equipment package into the lower stratosphere.

Their ideas seemed viable and in autumn 1946, the balloon project was classified as top secret and assigned the name Mogul. NYU were also asked to recruit a suitable scientist to head the programme and they provided Charles B. Moore (1920-2010). A secret contract (W28-099-ac-241) was then issued by Air Material Command (AMC) to NYU for the initial production of a constant-level balloon, and another classified contract (W28-499-ac-82) to Columbia University for the development of electronic detection equipment.

Mogul moved along swiftly and an initial group from NYU, headed by the geophysicist Albert P. Crary (1911-1987), set up the operation at Alamogordo AAF, New Mexico.

A typical Mogul balloon system consisted of 26 or 28 neoprene, helium-filled meteorological balloons linked by a nylon cord measuring (on average) about 600-675ft (183-205m). The balloon train's primary purpose was to carry the detection package

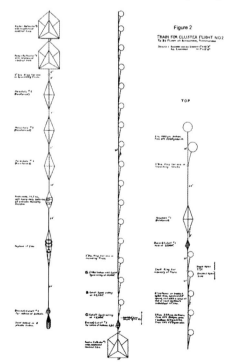

that utilised a sonobuoy microphone, amplifier, radio transmitter and power pack. In addition, the train lifted the automatic ballast system developed by Moore and three (lightweight) ML-307B Rawin radar targets made from aluminium foil, which would be used to supplement visual tracking.

Project Mogul test flights started at Alamogordo AAF (later Holloman AFB) in New Mexico during June and this series continued until July 1947, with launches following over the next three years and eventually totalling 110. The operational height these balloons were tested at is not entirely clear, but it was probably somewhere in the region of 45-50,000ft (13,700-15,000m). It is also known that the use of polythene gradually replaced neoprene as balloon material.

The project is very poorly documented and there are many gaps in the detail. To test the electronic equipment, some of the early flights involved the detonation of explosives within the vast nearby White Sands Missile Range (WSMR). In April and May 1948, Mogul balloons were flown (from several unspecified locations) during the Operation Sandstone nuclear weapons tests conducted at Eniwetok Atoll and it seems that some positive results were returned. The project continued in some capacity until early 1950, when it was officially closed. It has been said that Project Mogul was very expensive, difficult to operate and unreliable.

In fact, the first Soviet nuclear test was confirmed in September 1949, using specially adapted aircraft that gathered airborne radioactive particles. Exactly why Project Mogul was so highly classified remains a complete mystery and even the name Mogul was never known to any of the USAAF or NYU personnel involved with the programme.

THE MANTELL CASE

Some months after the Roswell Incident, another UFO event in the United States would generate headlines around the world.

At lunchtime on 7 January 1948, the police at Madisonville, Kentucky, began to receive calls from concerned members of the public who had sighted a large round brilliant object in the sky. Reports were soon coming in from residents at Irvington and Owensboro, and the Kentucky Air National Guard were alerted. There was now considerable concern about the presence of this unidentified object in the proximity of Fort Knox, which houses the US gold reserve.

At 1.45pm, the object could be seen from Godman AFB and personnel reported it to be slowly oscillating and changing colour from red to white. For about an hour the Commanding Officer, Colonel Guy Hix, observed the mysterious object with binoculars, later saying it was definitely not the planet Venus or any obviously identifiable object. He described the UFO as looking rather like an umbrella or parachute and sometimes resembling an ice cream cone topped with red.

At this point, the Godman Tower was contacted by a group of four Kentucky Air National Guard F-51D (originally designated P-51D) Mustang fighters. They had set out earlier from Marietta AFB in Georgia and were delivering the unarmed aircraft to new users at Standiford AFB, Kentucky. Leading this flight was Captain Thomas 'Tommy' Francis Mantell in aircraft KY-869 and Godman Tower asked him to investigate the object.

For various reasons involving fuel or oxygen, the other three Mustangs abandoned the interception, leaving Mantell to climb towards the UFO. Because this was a short-

Above: **Captain Thomas Mantell (centre), who lost his life intercepting a UFO over Kentucky in 1948.** *Bill Rose*

range ferry flight, the aircraft had not been provided with oxygen, necessary for high-altitude operations. However, Mantell continued to climb and as an experienced combat veteran, was aware of the risks posed by high-altitude operations.

As he closed on the object, Mantell reported to Godman Tower, "It appears to be a metallic object or possibly reflection of the Sun from a metallic object, and it is of tremendous size." He continued to climb, ignoring regulations by exceeding an altitude of 14,000ft (4,267m) without oxygen, and soon passed 22,000ft (6,705m). Then radio contact was lost and minutes later his fighter dropped from the sky and ploughed into the ground on Joe Phillips's farm, about 3 miles (4.8km) south-west of Franklin, Kentucky.

In the weeks that followed, a series of

Below: **Reconstruction of the F-51D Mustang fighter flown by Thomas Mantell of the Kentucky Air National Guard during 1948.** *Bill Rose*

sensational reports appeared in the tabloid press. Many suggested that the mysterious object was a huge alien spacecraft and Mantell had been shot down with a death ray. Other stories claimed that his body was missing from the wreckage, or had been strangely disfigured. None of this was true, although there may have been another cover-up to hide the truth, which involved a military balloon.

It is most likely that that Mantell lost consciousness due to anoxia (oxygen starvation) when his aircraft reached an altitude of around 25,000ft (7,620m). The Mustang probably continued to climb to about 30,000ft (9,144m) where it briefly levelled off and then plunged into a steep dive. As for the UFO, it was almost certainly a large Skyhook balloon. It has been suggested that this was launched from Clinton County Air Base (now a commercial site) near Wilmington, Ohio, but records indicate that there were no Skyhook releases from this location before July 1951.

According to Professor Charles B. Moore, who went on from Mogul to head the US Navy's Skyhook programme, this particular balloon was almost certainly launched from Camp Ripley, Minnesota, on 6 January 1948. Moore claimed that it carried cosmic ray detection equipment and nothing sensitive, but suggested that the US Navy was reluctant to admit it had played any part in the death of a flyer, and personnel were immediately instructed not to discuss the matter.

Winds carried this Skyhook balloon across Kentucky at an estimated altitude of around 70,000ft (21,336m), with weather conditions allowing widespread sightings from the ground. As it moved and shifted shape slightly, the reflected sunlight would sometimes give the balloon's envelope a red appearance and in this case the effect was quite dramatic.

Balloons created many UFO sightings, with the USAF eventually claiming that during the time they conducted investigations, the figure for this was at least 21.3 percent. While the Mantell case has become another cornerstone of UFO folklore, all available evidence does seem to indicate that this really was nothing more than an unusual balloon sighting linked directly to a tragedy.

SKYHOOK

Skyhook began in the immediate post-war years and was outwardly a scientific balloon programme operated by the US Navy. The USAF also launched a number of Skyhook

Above: **A Skyhook balloon is filled with helium prior to launch from the US Navy aircraft carrier** *Valley Forge* **in 1960.** *US Navy*

Right: **A huge helium-filled Skyhook balloon is prepared for launch from the deck of the US Navy aircraft carrier** *Valley Forge*, **located about 100 miles (160km) south of the Virgin Islands. This was one of several launches undertaken in January 1960 as part of Operation Skyhook, which was officially a scientific endeavour.** *US Navy*

balloons for research purposes, with both services making flights until about 1976. Many launches were made from locations within the United States and from vessels at sea.

The most important feature of the Skyhook balloon was its polythene envelope, manufactured by the Aeronautical Research Division of General Mills, a company better known for food processing. Polythene was far superior to neoprene in every respect. The

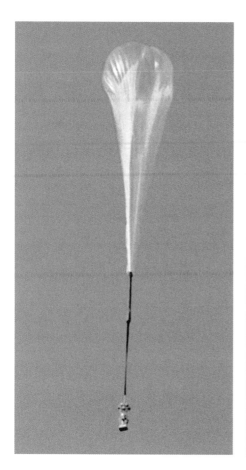

material weighed less, retained helium for longer, and was more resistant to temperature extremes, ozone and the effects of ultra-violet radiation. It was also cheaper to produce and easier to work with.

Charles Moore was largely responsible for the development of these new balloons. He was immediately aware that it would be possible to routinely lift useful payload into the upper levels of the stratosphere, where they would remain for several days. Generally, the average US Navy Skyhook balloon had an approximate length of 75ft (22.9m) and a diameter of about 100ft (30.5m). For certain specialised missions, many large balloons were fabricated with lengths in excess of 400ft (122m) and diameters of 300ft (91m).

When atmospheric conditions were good, the significant size of Skyhook balloons could make them visible at an altitude of 100,000ft (30,500m) or even higher to an observer on the ground. Furthermore, the observable slant range would often be in excess of 85 miles (136km). Skyhook initially operated alongside Project Mogul and because of this programme's success, the complex Mogul balloon trains eventually gave way to polythene enclosures.

Scientific missions undertaken by

Skyhook balloons included cosmic ray detection and specialised photography. One particularly interesting example was the 1957 high-altitude Stratoscope Project, which provided the first high quality images of sunspots using a 12in (30.5cm) telescope. While Skyhook balloons were ostensibly being used for meteorological and scientific research purposes, it was clear that there was potential for specialised military and covert operations. For this reason, Skyhook began to move into classified territory.

Some Skyhook launches, especially those undertaken by the USAAF/USAF from Alamogordo Army Airfield (later Holloman AFB), New Mexico, appear to have carried highly classified payloads. The kind of equipment they lifted was almost certainly prototype camera systems for high-altitude spyplanes such as the Lockheed U-2, electronic eavesdropping technology and experimental particle sampling equipment that would evolve into the Ash Can and Grab Bag balloon projects. Apparently, some of the larger Skyhook balloons launched at this time lifted unspecified payloads in excess of five tons to altitudes above 100,000ft (30,500m).

One known development was inspired by the wartime Japanese Fu-Go balloon bomb.

Above left: **A US Navy Skyhook balloon shortly after launch.** *US Navy*

Above: **Skyhook balloon No 93 leaving the deck of USS** *Norton Sound* **(AV-11) during early 1949.** *US Navy*

Below: **Seen from the ground, Skyhook balloons often generated reports of circular silver-coloured UFOs.** *US Navy*

Left: **During the 1950s, there were several classified US programmes that used unmanned high-altitude balloons to photograph areas of interest within the Soviet Union. The balloons flew above the reach of Russian fighters and were unstoppable, but their use eventually led to a major diplomatic row.** *Bill Rose*

The history of this weapon is unclear, but by 1952, it had evolved into Weapon System 124A (WS-124A), codenamed Flying Cloud. Using inexpensive hydrogen as a lift-gas, it would be used to spread chemical and biological agents, with the possible emphasis on destroying crops. Tests proved that the concept was very unreliable and it was abandoned at the end of 1954. Whether the idea of using a large balloon to disperse radiological material was ever considered is unknown. Yet another early 1950s project was given the name Moby Dick (after Herman Melville's novel of 1851) and this was used to map high-altitude winds.

Starting in the late 1940s, numerous large balloons were released by the US Navy and USAF in many locations across the United States, with some balloons drifting across the entire country. Sometimes slow moving, the large polythene envelopes would lose their shape, occasionally becoming oval, spherical or disc-like. The balloons would

also sometimes reflect sunlight not visible from the ground and appear to change colour.

With elevated public interest in UFOs, it is hardly surprising that so many chance sightings of Skyhook balloons were reported as UFOs and knowledge of them was denied due to their sensitive payloads.

GENETRIX

A study by the RAND Corporation in 1951 had concluded that Skyhook balloons, although uncontrollable after launch, might be used for reconnaissance missions above the Soviet Union and China.

The Cold War had triggered a deadly East-West arms race and Washington was desperate for information on military developments behind the Iron Curtain. Reconnaissance flights using manned aircraft were largely out of the question and orbital space systems remained something for the future. On the other hand, balloon technology had advanced considerably and although far from perfect, it was felt that a mass launch might return some useful results. Furthermore, it is possible that Washington was perfectly willing to anger the Russians with such an operation. After considerable deliberation, President Eisenhower authorised this highly classified CIA sponsored programme on 27 December 1955.

Known as Project Genetrix, the idea was to launch hundreds of camera-equipped balloons that would cross the Soviet Union and China at a constant altitude of 72,000ft (21,946m). The programme was organised by the USAF and it is obvious that Project Genetrix had been in preparation for some time. On 10 January 1956, weather conditions were considered good enough to begin the operation and the first 9 polythene balloons were launched from the USAF base at Incirlik, Turkey. The project now rapidly gained momentum with further launches being made from Gardermoen in Norway, Evanton in Scotland and two different locations in West Germany. It is believed that some were also released by the US Navy from aircraft carriers.

The balloons would maintain a constant height of 72,000ft (21,945m), placing them beyond the reach of Russian air defences. They would drift across the continent for several days, taking hundreds of pictures during daylight hours until they reached the retrieval area. At this location, a specially modified C-119F transport aircraft operating from Japan or Alaska would be waiting. An

encrypted radio signal would then be sent to the balloon's gondola releasing it by parachute and the C-119F would perform a mid-air recovery. This proved to be quite a dangerous undertaking, but every payload was potentially of high value. About halfway through the project, Pentagon officials took the seemingly irrational decision to lower the operational height of these balloons to 55,000ft (16,764m), making it much easier for the Soviets to take action against them.

The Russians were now fully aware of what was taking place and although the balloons were difficult to track with radar, they were being continually reported in daylight or moonlight by ground-based observers. At the reduced operational altitude, it was now possible for Russian fighters to shoot down a few balloons and the camera packages were subsequently put on public display.

This situation now came to a head on 4 February 1956, when Soviet Deputy Foreign Minister Andrei Gromyko presented a formal protest to the US Ambassador in Moscow. With tensions rising, President Eisenhower ordered a halt to the programme two days later. Although the launches ceased, Genetrix recovery continued until the start of March.

When Genetrix ended, about 446 balloons had been released and around 40 gondolas recovered. (Some sources indicate a handful more). Of these, some camera units malfunctioned, leaving about 34 useful packages of exposed film. The value of Genetrix remains debatable. It was expensive, generated major political problems and returned limited information about Russia and China. Perhaps the only major discovery was the Dodonovo nuclear processing plant in Central Siberia.

MELTING POT

Although it had been agreed to stop launching balloons on flights across the USSR, the CIA remained eager to gather information on military developments within the communist countries using any means available. One approach was sponsorship of a high-altitude spyplane called the Lockheed U-2, and the other was a further development of Skyhook called WS-461L.

This balloon system was designed to operate at 100,000ft (30,480m) which would make it very hard to detect and put it well beyond the reach of any foreseeable methods of interception. Able to maintain its height and circle the Earth in a month, WS-461L carried advanced photo-reconnaissance equipment. This was primarily a state-of-the-art Itek HYAC-1 large-format roll-film panoramic camera designed by the company's chief engineer, Frank Madden, and capable of recording very high resolution images.

There were growing concerns about the development of long-range Russian missiles and this convinced President Eisenhower to approve the secret use of the second balloon system on 25 June. This was given the project name Melting Pot and it would be undertaken by the US Navy with some support by USAF specialists. Having completed a number of trials, the first launches were made from the US aircraft carrier *Windham Bay* in the Bering Sea. This location was chosen because of unusual high-altitude air currents.

Starting on 2 July 1958, a total of 7 WS-461-L balloons were launched which carried simple meteorological equipment to mask the true purpose of the operation. One of these failed, then the highly classified balloons (S-430, 431, 432) were released which carried the Itek cameras. The operation concluded on 14 July and the USS *Windham Bay* returned to port.

All went well until the balloons reached Poland and the gondolas were accidentally released, handing America's most advanced reconnaissance equipment to the Soviets. It also generated another major diplomatic incident and spelt the end of all further covert balloon projects. However, there is some evidence that the Americans continued to undertake occasional reconnaissance flights using balloons, and claims emerged from Russia that they were still being shot down during the mid-1970s. Assuming another US reconnaissance balloon programme followed Melting Pot, it remains classified.

I hope the last sections have allowed me to show the importance of military balloons in the immediate post-war years, the amount of secrecy involved with some specialised military projects, and the substantial link to UFO/flying saucer sightings.

GOODYEAR CIRCULAR BALLOON

Today, helium and hot air balloons can be manufactured to have the appearance of a flying saucer. An interesting early example of this shape was designed by Robert Ross and Jean Charpentier of Goodyear Aerospace in the late 1960s. Intended for commercial or military applications, the key feature of this helium-filled wing balloon was its ability to maintain a constant attitude by presenting the same profile to wind coming from any direction. Secured to the ground by cables, this type of balloon can be used for a wide range of purposes.

DENTON LTA CRAFT

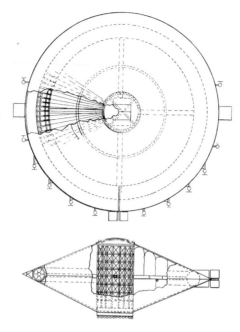

In the late 1960s, packaging engineer and designer Harvey R. Denton (1901-1974) turned his attention to developing a circular lighter-than-air craft capable of carrying very substantial payloads. The overall shape resembled a flying saucer, which was formed from a rigid framework covered with a metal skin. Gas bags filled with helium would occupy most of the craft's interior.

A substantial central core made from

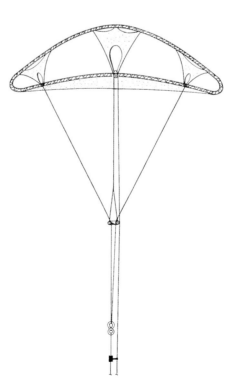

aluminium alloy supported a lower section for the crew, passengers and cargo. On the ground, this section was supported by a series of wheels. Small propellers would be used for manoeuvring during take-off and landing or for horizontal flight, and these would be driven by electric motors powered by diesel engines or nuclear power. A small tail unit would assist with flight control.

Denton seems to have been confident that his design would generate considerable interest and it was filed as a US patent on 12 March 1973, but he died the following year, before it was published (3820744) and the design was largely forgotten about.

MacNEILL LTA SAUCER

This design for a large lighter-than-air flying saucer was completed in the mid-1970s by Roderick M. MacNeill (1918-2000), an American engineer who hoped to develop a low-cost transport aircraft capable of carrying passengers or cargo.

His flying disc would be built from a lightweight alloy framework and covered by a metal or composite skin. In the upper centre of the aircraft, a large transparent dome would enclose the flight crew and passengers, with an elevator unit connecting to a lower cargo bay area. Most of the aircraft's interior would be filled with modules containing helium lift-gas and the aircraft would effectively have a full VTOL capability.

The estimated diameter of MacNeill's aircraft was about 160-180ft (48-54m) and the overall height would be approximately 35ft (10m). Two different propulsion systems were proposed. MacNeill's first idea was to utilise two turbofan engines in swivelling wingtip pods for direct ascent and horizontal flight. His second idea was to use a ducted fan system that channelled the airflow for either lift of horizontal flight. Control of the aircraft was undertaken by orientating the jet engines to provide roll, pitch and yaw. Alternatively, the ducted fan system could be used selectively. Another feature would be a series of automatically controlled ailerons around the periphery of the disc.

MacNeill's objective was to minimise drag and wind resistance and provide a cost-effective transportation system that would eliminate the need for normal airport facilities. His design was filed as a US patent on 15 September 1975, and was published on 29 March 1977, but it failed to attract any commercial interest.

Flying Saucer Technology

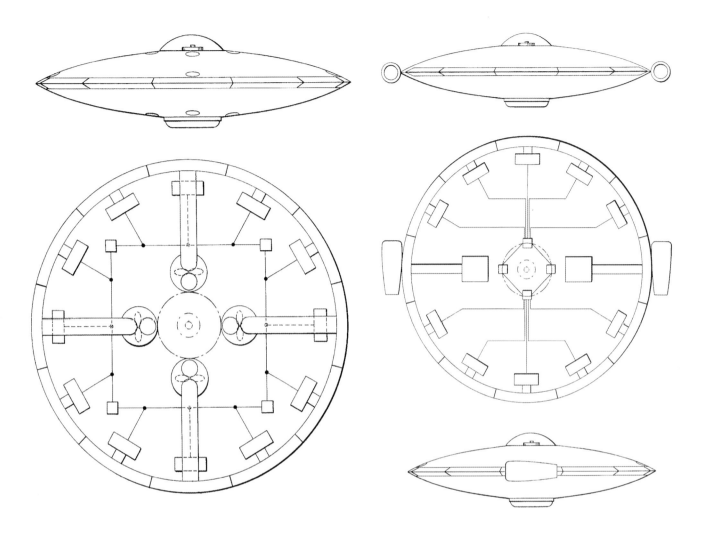

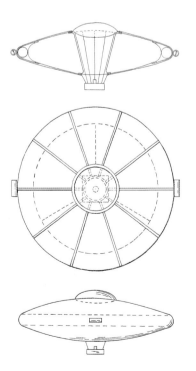

Above left: **The MacNeill hybrid flying saucer equipped with wingtip-mounted gas turbine engines.** *Bill Rose*

Above right: **This version of the MacNeill flying saucer is propelled by four ducted fan units providing lift for VTOL operations and level flight.** *Bill Rose*

Left: **This non-rigid saucer-shaped balloon was designed by the solar-powered balloon pioneer Fredrick Eshoo.** *Bill Rose*

ESHOO NON-RIGID SAUCER

During the 1970s, the well-known Iranian balloonist Fredrick Eshoo (1940-) turned his attention to the idea of designing a non-rigid flying saucer shaped aircraft that would be partly filled with helium and also use heated air.

His primary objective was to demonstrate that the saucer shape could be maintained in level flight. It was also hoped to reduce the level of drag associated with conventional manned balloons.

The gondola carrying the pilot is located directly below the centre of the airship and is suspended by load cables. The gondola also carries a propane burner and the centre section of this design is effectively a hot air balloon. In the surrounding exterior section of the airship are a number of separate chambers, and an annular tube circles the periphery. All contain helium to provide lift. A further feature considered necessary for this concept are two small electrically powered ducted fans used for propulsion and manoeuvring. An alternative proposal is equipped with a single pusher propeller mounted behind the airship, which is integral with a stabilising fin. Apparently, this design was aimed at the sports market but does not appear to have been built. Nevertheless, the idea would seem to be perfectly viable.

SKYSHIP

In May 1974, John West Design Associates based in Surrey, England, released details of a commercial airship project called Skyship. The idea originated with John West who worked as a naval architect with P&O and was largely responsible for the liner SS *Canberra*. West believed that a low-drag, lenticular-shaped craft with a VTOL capability would be ideal for transporting heavy loads to inaccessible areas. Such an airship, especially if inexpensive to manufacture, would be ideal for use by developing countries and of particular interest to the military. Assisting West with this project were Dr Cyril Laming and former Army Major Malcolm Wren, who were both engineers.

The aircraft would have a flying saucer shape, with a substantial diameter of 700ft (213m) and a maximum thickness of 203ft (62m). Its gross weight was set at 725 metric tonnes and ten ducted Rolls-Royce RB.109 Tyne turboprop engines provided propulsion. Skyship would cruise at 100mph (160kph) while maintaining an altitude of approximately 5,000ft (1,524m) and it would be possible to carry 360 metric tonnes (fuel and payload) over a distance of 1,000 miles (1,600km).

Much of the internal space would be occupied by modules containing helium lift-gas and West envisaged a crew of twenty-four. Skyship would be constructed from an aluminium alloy framework covered in fabric. West had intended to use an alloy skin, but the cost proved prohibitive and his objective was to make Skyship as affordable as possible.

When Skyship was announced in London,

the initial designs had been completed and the company was discussing plans to build a small prototype. One of the first specialists to become involved with the project was engineer and college lecturer John Gibbs, who was recruited by the company's Technical Manager Robin Wren. His first assignment was to build a small, accurately detailed desktop model of the aircraft for use during development reviews.

The model was formed from a ring beam connected to a central hub by a series of spokes. On the full-sized aircraft, the hub section would house the ducted fan propulsion system. The outer shell of the model was fabricated from fibreglass, showing smaller ducted fan ports. Not quite a true flying saucer, the model was equipped with a tail extension that could be fitted with control surfaces for pitch and roll. Following

a detailed review, it then was decided to build a proof of concept flying demonstrator. Robin Wren undertook the design work to West's specifications and Gibbs was retained as the construction engineer.

The helium-filled Skyship Demonstrator would have a diameter of 30ft (10m) and a height of 9ft 6in (2.9m). It was electrically powered and flown by remote control. Gibbs produced a ring beam using twelve equal sections made from 1/8in (3mm) plywood, which would be bolted together. Weight was of paramount importance and every section was lightened as far as possible. Spinnaker canvas was used to cover the craft's exterior and suitable material for the gas bags (thought to be Mylar) was sourced from a US supplier. It was then decided to use a mixture of hydrogen and helium, which was cheaper than pure helium and slightly more buoyant. For propulsion, four small electric motors driving nylon Keil Kraft 12 x 4 propellers were evenly positioned around the ring beam and several small 6-volt lead-acid motorcycle batteries provided the electrical power. The equipment payload weighed 40lb (18kg) and the lift weight was 215lb (97.5kg).

The motors were controlled by a modified Flight Link multi-channel proportional radio control system, allowing full directional control.

Manufacture of the demonstrator's components was completed by the end of 1974 and assembly followed inside the Airship Hangar at Cardington, Bedfordshire. This enormous building had been erected in

Top: **The completed Skyship Demonstrator within the huge Airship Hangar at Cardington prior to flight trials.** *John Gibbs*

Middle:**Testing the Skyship Demonstrator at Cardington in 1975.** *Bill Rose Collection*

Bottom: **Model of the production Skyship, shown tethered at a ground station.** *John Gibbs*

1930 to house the R101 airship and was considered the ideal location to launch a new British airship project.

Assembly was completed in April 1975 and neutral buoyancy was achieved without problems. The next stage was flying trials within the hangar and the media were invited to observe developments, with the BBC's *Tomorrow's World* programme taking a special interest in the project. However, there were problems controlling the vehicle, which Robin Wren eventually traced to interference with the radio control link caused by the hangar's overhead fluorescent lighting.

It was hoped that these test flights would lead directly to the construction of a manned prototype with a 200ft (61m) diameter and a 9 metric tonne payload capability, but there was insufficient commercial interest to proceed.

Consequently, a new and more capable hybrid design was produced, which Malcolm Wren called Thermo-Skyship. This was similar in appearance to Skyship, utilising helium-filled modules to provide neutral buoyancy but also employing heated air or steam within the craft to generate lift.

In March 1976, Major Wren tried to interest the Ministry of Defence (MoD) in sponsoring development of the Thermo-Skyship as a troop transporter and although there was interest in the project, the MoD were unwilling to provide financial support. As a consequence, Major Wren returned to seeking a commercial sponsor and, in June 1978, he established a new company called Thermo-Skyships Ltd based at Ramsey, Isle of Man. Wren also caught the attention of a major UK ferry company which was considering the possibility of operating city centre airship services between London, Paris, Brussels, Amsterdam and Düsseldorf.

Thermo-Skyships were proposing an initial flying saucer shaped prototype called TS-26, which would be capable of carrying a 5,000lb (2,267kg) payload over a 1,250 mile (2,000km) distance. This would be followed by a larger production craft with a diameter of 210ft (64m), a payload capacity of 10 metric tonnes (or 100 passengers) and a range of 450 miles (725km). The estimated cruise speed would be 100mph (160kph).

At this point in time Thermo-Skyship's Deputy Managing Director David Potter was suggesting that a prototype could be completed by 1980, with production of the larger airship beginning two years later. He also claimed that it would be feasible to build a very large version of the airship by 1990, with the ability to carry a 150 metric tonne payload or 1000 passengers.

Throughout 1979, Thermo-Skyships continued to recruit designers and engineers to work at their Ramsey offices, but there was still a lack of serious commercial interest in the project. The following year, Thermo-Skyships took over Airship Developments Ltd, to become Airship Industries, and in July 1980, Redcoat Airlines based at Gatwick agreed to purchase four freight-carrying airships from the company. The history of these organisations is rather hard to follow but by 1982, another company operating from the same Ramsey address had come into existence. It was run by Major Malcolm Wren and was called Wren Skyships.

Representatives of Wren Skyships attempted to sell the Thermo-Skyship design to customers in the Middle East and New Zealand. Major Wren also investigated the idea of building airships at Youngstown, Ohio, for the American market and examined the possibility of operating an airship service between the Isle of Man and Blackpool. Following this, there were financial difficulties and Wren Skyships ceased trading by the start of the 1990s.

NEUMAYR'S LONG-RANGE SAUCER TRANSPORT

After retiring from Bell Aerospace, George A. Neumayr (1925-2004) relocated to Florida. He had worked for the company as a design engineer and now spent much of his time developing a VTOL lighter-than-air flying disc concept. Neumayr was a fan of SF movies and intrigued by flying saucers, which encouraged him to pursue this kind of design. He wanted to create a VTOL aircraft of substantial size that would eliminate the need for runways, reduce wind shear, air pocket drops and icing problems.

The idea began to form in about 1970, and he envisaged a lenticular-shaped vehicle that would use most of its upper internal space to accommodate a lift-gas such as helium.

As the concept moved towards a more detailed study, he decided that the craft would have a diameter of 200ft (60.9m) and a height of 50ft (15m). The upper surface of the aircraft would be covered with solar panels to supply power for many of the internal systems. Passengers, cargo, propulsive units and fuel would be located in the aircraft's lower section. Fitted with large observation windows, the craft would accommodate at least 750 passengers.

It would employ two or four turbofan engines for horizontal flight and eight internal

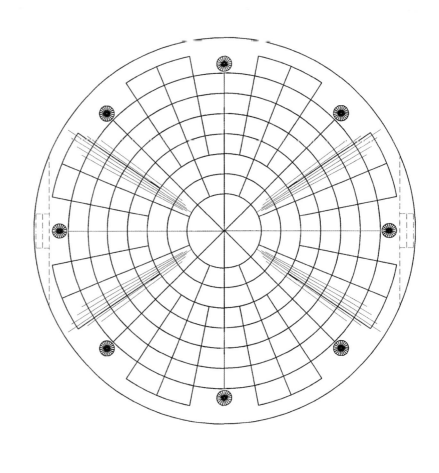

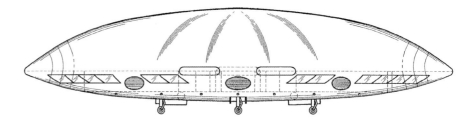

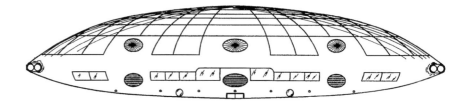

Top: **Former Bell Aerospace engineer George Neumayr designed this large, hybrid flying disc, which he expected to rival a jumbo jet in terms of performance.** *US Patent Office/Bill Rose*

Bottom: **Forward view of the Neumayr flying saucer concept.** *US Patent Office/Bill Rose*

Flying Saucer Technology

ducted gas turbines dispersed around the disc to facilitate lift-off, hover and landing. Each duct outlet would be fitted with a louvred shutter unit that could be rotated to provide control of pitch, roll and yaw. The aircraft's main structure would be built from aluminium or magnesium alloy and be covered in a solid skin. Neumayr was confident that it would be extremely strong and even capable of floating on water. While on the ground, the aircraft would be supported by a four-strut, eight-wheel fully retractable undercarriage.

Although Neumayr confidently predicted that his aircraft would have comparable performance to a jumbo jet and enjoy lower noise levels, there are some unexplained grey areas with this proposal, such as the aircraft's anticipated operational altitude. The likelihood of any major aerospace company taking a serious interest in such a technically challenging design was always slight. Nevertheless, Neumayr applied for a US patent (5351911, published 4 October 1994) and forwarded details to many companies and government departments. There were no takers.

HENRY'S DISCOID AIRCRAFT

This compact VTOL disc-shaped airship was designed by American engineer Roy Henry in the 1990s. The main section contains helium-filled modules for lift, with the objective being neutral buoyancy.

Propulsion is provided by a centrally located gas turbine engine in an upright position, which channels its exhaust into a series of valve-controlled ducts with outlet nozzles positioned in four equally spaced groups around the circumference of the disc (front, rear and sides). These are used to provide lift, directional control and horizontal flight. Directly below the engine is the gondola containing the pilot. The engine draws air from an inlet that surrounds the top of the gondola and the aircraft is supported on the ground by a fixed position tricycle undercarriage.

It is not entirely clear what Henry was aiming for with this design, and the aircraft does not appear to have found the necessary financial backing for further development.

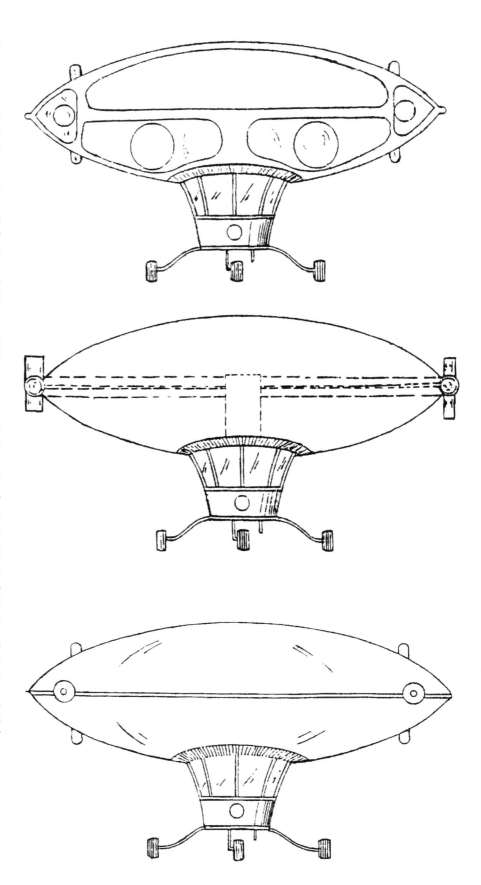

Top and Middle: **Forward view of Roy Henry's hybrid discoid aircraft proposal. The lower drawing shows the position of the propulsion and duct system.** *US Patent Office*

Bottom: **Forward view of Roy Henry's hybrid discoid aircraft proposal.** *US Patent Office*

Above: **The 131ft (40m) diameter prototype Aviastar ALA-40 Thermoplane undergoing ground tests. Utilising a Mil Mi-2 helicopter fuselage, it was hoped that the ALA-40 would lead to a much larger 650ft (200m) diameter craft with a 500-600 ton payload capability.** *Aviastar*

Left: **This drawing shows one of the many configurations studied for a large, heavy-lift Aviastar Thermoplane airship.** *Aviastar*

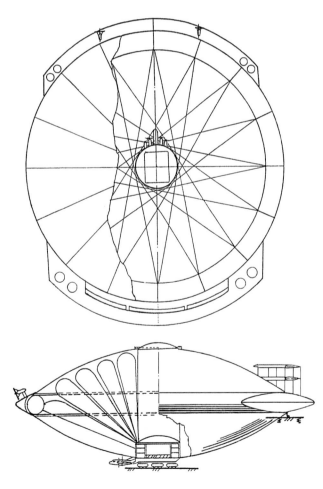

AVIASTAR THERMOPLANE ALA-40

Work on a State-funded 'two-gas' airship started in the Soviet Union during 1985 under the direction of the designer Yuri Ishkov. His objective was to produce a low-cost, rigid structure, lenticular-shaped aircraft with superior aerodynamic features, a heavy lift capability and VTOL capability. This was considered essential for use in inaccessible and remote areas. An important feature of the concept was the 'two-gas' system. The aircraft would be fitted with modules containing a lift-gas such as helium and, in addition, hot air would be generated in a system not too dissimilar to the British Thermo-Skyship.

In 1989, a decision was taken to build three prototypes and the airship was named Thermoplane. However, the Soviet economy was in serious trouble and funding for the project became increasingly uncertain. Nevertheless, the Aviastar factory in Ulyanovsk still managed to produce the first ALA-40 01 prototype in 1992 and ground tests began within a matter of months.

Above: **Designed by Yuri Ishkov, the prototype rigid structure Aviastar ALA-40 airship during initial trials at Ulyanovsk, Russia, in 1992.** *Aviastar*

ALA-40 01 was 131ft (40m) in diameter and capable of lifting a modest 5-6 ton payload. A rear stabiliser with rudders was fitted to the aircraft and small propellers were used for forward flight and manoeuvring. The cabin compartment was built from the fuselage of a surplus Mil Mi-2 helicopter. The first tethered flight of the Thermoplane was undertaken in 1993, and two further prototypes were in the early stages of construction with completion of ALA-40 02 expected in 1995.

A number of larger versions were considered, the most promising design being known as ALA-600. This had a diameter of 770ft (235m) and the ability to lift 500-600 metric tonne payloads. It was expected to have a maximum level speed of about 130mph (210kph) and an estimated range of 3,000 miles (5,000km). Even larger aircraft with a diameter of 1,000ft (320m) and very large payload capabilities were considered feasible.

Unfortunately, Aviastar were unable to secure any further funding from the State, or attract any outside commercial interest, so

ALA-40 01 was mothballed. There were discussions with a Middle Eastern company in around 2000 to complete a modified version of the 02 prototype, but this came to nothing and the Thermoplane project remains mothballed.

THE CHRYSALIS SAUCER HOAX

To date, the most elaborate (known) flying saucer hoax took place on 16 August 2003 and was orchestrated by a British TV production company. The stunt was filmed and shown on 7 October 2003 by Channel 4 as *A Very British UFO Hoax*. The idea originated with Mark Raphael, an executive producer with Chrysalis TV who had been considering a number of ideas for a fake UFO event. It is believed that he was inspired by Sir Richard Branson's highly publicised 1989 April Fools' Day stunt, which made use of a hot air balloon that was designed to look like an alien spacecraft.

Raphael concluded that the best approach was to construct a convincing remote-controlled flying saucer and fly it across the sky in the presence of UFO skywatchers while secretly recording their reactions. The plan evolved into a detailed proposal and Channel 4 TV was approached in late 2002 to

sponsor the project. The programme would be made as a documentary to fill a 60-minute slot and Raphael suggested that the working title should be 'How to build a flying saucer'.

Channel 4 executives were impressed with Raphael's proposal and decided to commission the project. In turn, Chrysalis contracted Cutting Edge Effects Ltd to design, build and operate a convincing remote-controlled flying saucer model. Cutting Edge were a leading UK special effects company based at Elstree Studios, Borehamwood, who had worked on many big budget movies such as the Bond film *Goldeneye* (1995).

They had already expressed an interest in the project, deciding it was technically achievable and affordable, and a large model would convince observers on the ground that they were looking at a genuine UFO. Supplying the expertise for this flying saucer project were Robbie Scott and Nigel Blake. They had apparently quoted Raphael a figure of around £50,000 to develop the flying saucer, which would take about eight months to complete.

The design for the flying saucer was left to Scott and Blake, who reviewed hundreds of images from books and magazines before submitting their final proposal to Raphael.

In appearance their model would be very

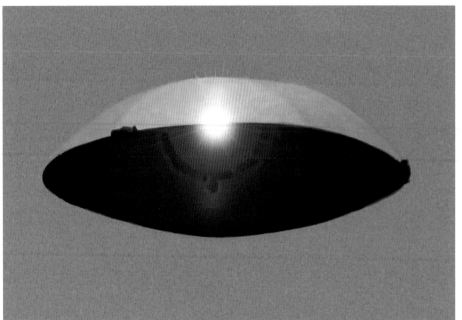

Right: **A re-creation of the Chrysalis flying saucer, built by film industry special effects company Cutting Edge Effects for the 2003 UK TV production** *A Very British UFO Hoax*. Bill Rose

similar to the John West Skyship Demonstrator built by John Gibbs in 1975. Not surprisingly, the choice of constructional materials was rather different and the slightly smaller ring beam would be made from carbon fibre in eight separate sections. Two helium-filled compartments would provide lift and a lightweight reflective material was chosen for the model's skin to create the illusion of a solid metallic object. When completed, the Chrysalis saucer would have a diameter of about 25ft (7.62m) and a central thickness of approximately 8-9ft (2.4-2.7m). Cameron Balloons Ltd, based in Bristol, were asked to provide suitable material to cover the craft and they supplied 984ft (300m) of helium fabric composite (with a silver Mylar laminate) which was assembled using Cameron's sewn/sealed technique.

Experiments to cover the saucer's framework with this material were organised by Cameron's Gas Department Manager Gavin Hayles. This went fairly well, although there were concerns about the flying saucer's appearance caused by wrinkles in the Mylar, as it was essential that the saucer appeared to be made from smooth metal. Problems also arose with the small German-made fans that proved to be incompatible with the ducted propulsion units and this was followed by a delay obtaining the model's American-made 5kW power source, which prompted Scott to collect it from California.

However, the project seemed to be in serious trouble when Cameron Balloons delivered the silver helium skin in late July 2003 and it was found that the model was overloaded by about 24lb (10.9kg). The only

way to deal with this issue was to reduce the weight further by drilling holes in the carbon fibre hoop and dropping plans to use the generator. This was replaced with nickel-cadmium power packs wired to each motor. The problem had been solved and it would still be possible to attain a maximum speed of 20mph (32kph) and fly the model for twenty minutes before the battery packs were completely exhausted.

While this change was unwelcome, there would be just enough time to complete the proposed flight if everything went according to plan. The last component to be installed on the central equipment platform was a radio control unit, linked to each ducted fan unit via a network of fibre optic cables. When the envelope was inflated, this could be accessed via a small tunnel.

Preparations

The model components were now moved to a large hangar at RAF Aston Down, which is a largely disused World War 2 military airfield situated between Stroud and Cirencester. Once assembled, neutral buoyancy was achieved and the saucer was ready for testing.

Cutting Edge had hired the well-known model aircraft flyer John Palmer, who took charge of flight operations and quickly determined that at least seven pilots would be required to guide the model across the planned distance. Unfortunately, the hangar trials ran into difficulties and Palmer found it almost impossible to control pitch during early tethered flights. This was eventually

rectified by the attachment of a small cardboard spoiler beneath the model, which was then replaced by a more refined version covered in reflective film. The next requirement was to obtain certification from the Civil Aviation Authority (CAA) which was represented by John Greenfield, a respected name in the model aircraft field.

After several demonstrations, Greenfield was satisfied that the model was airworthy and could be safely controlled in flight by all the chosen operators. Consequently, he issued an Air Navigation Order and Exemption Certificate, which limited the model's ceiling to 400ft (122m) and specified that the visual control range should not exceed 1,640ft (500m). He also insisted on the use of an automatic fail-safe system, which would force the vehicle to land in the event of a control system failure. Finally, it was agreed that the local air traffic controllers would be advised about the flight prior to it taking place.

As the test phase progressed, Chrysalis Producer Chris Harries and Robbie Scott completed their survey of suitable locations to undertake the hoax. At the top of their list was an area of Rendlesham Forest in Suffolk, near to the now-defunct USAF Bentwaters base. This was the scene of a very controversial UFO event on 27 December 1980, which many members of the UFO community still regard as Britain's equivalent to the Roswell Incident.

It appeared to be the obvious choice, but there were major problems with this location, including access, anticipated visibility of the model by sufficient numbers of unprepared observers and flying restrictions within military airspace. So Rendlesham was abandoned and a decision was made to use the second choice – Avebury, Wiltshire – just a few miles north of Salisbury Plain. This historic village is famous for its mysterious stone circle dating back to around 2,500BC and this area has a reputation for paranormal activity, crop circles and UFO sightings.

In June 2003, Harries contacted Brian James, a well-known UFO investigator. James is a former chairman of the British UFO Research Association (BUFORA) and currently heads a group of UFO investigators called the Anomalous Phenomena Research Agency (APRA). Harries said that he had been in touch with several other UFO organisations

Flying Saucer Technology

Above: **The Chrysalis flying saucer during flight-testing. The Cutting Edge Effects team are lined up beneath the model.** *Nigel Blake*

and wondered if APRA would be interested in contributing to a future TV documentary called *The Believers*. James later told the author that, in his opinion, the only purpose of this discussion was to confirm when members of APRA would meet at the 'Red Lion' pub in Avebury before moving on to undertake one of their UFO skywatches.

The exact details of what followed are not entirely clear, but Harries had learned that the APRA skywatch party would take place on 16 August. A production meeting followed and it appears that based on this information, a decision was taken to make the flight at that location and capitalise on APRA's presence.

The Flight

On 16 August, the Chrysalis team gathered in a secluded farmer's field, approximately 2 miles (3.2km) south-east of Avebury and just beyond the Ridgeway near Overton Hill.

Their activities were fairly well hidden from any chance observers, but the model was inflated with helium beneath a canopy while the camera crews and flight controllers were preparing for the stunt. Everything was proceeding well, until the wind changed direction from predominantly south-westerly to easterly and was picking up speed. With a

limit of 5mph (8kph) imposed by the CAA, there was a real possibility that the flight would have to be abandoned.

Robbie Scott now made some changes to the flight path, allowing the saucer to pass closer to Avebury, which would considerably reduce the flight time. However, it also meant that the lift-off point would be closer to the village. With the saucer inflated and all systems fully operational, members of the team carefully moved it across the fields to the new launch site. There was little chance of being seen, but the saucer was highly reflective, so it was covered with military camouflage netting during transit.

Everything was in place to begin the stunt. The wind speed was manageable, members of APRA had arrived at the pub, the camera crews were ready and twilight was approaching. The saucer now lifted into the early evening sky under the control of world champion model aircraft pilot Steve Elias. It would be steered towards the village, briefly hover above a flat area of ground and then turn north to a concealed recovery point, accessed from the Winterbourne Monkton road.

Everything was going to plan as the saucer steadily flew towards Avebury at a height of about 200ft (61m). Control was passed to

Steve Ansell and then John Palmer who were both in hidden positions along the Wessex Ridgeway. Sitting outside the 'Red Lion' pub were four unsuspecting APRA skywatch members: James Hill, Tim Field, Jason Hawkes and Brian James. As the flying saucer model came into sight at around 8.50pm, a member of the Chrysalis team arrived outside the pub and attempted to draw everyone's attention to the mystery object.

In fact, the APRA members were less than impressed by the craft and immediately determined that the flying saucer was some kind of a balloon that struggled to remain steady in the breeze. They also said that the noise of the craft's electric motors could easily be heard. Brian James later told the author, "At no point were we remotely fooled by this remote-controlled balloon, but I do admit that we (APRA) were guilty of dismissing the event as a student prank rather than a highly organised and expensive stunt devised and executed by a professional film company."

While the pub's customers stared at the flying saucer, a film crew using amateur

digital camera equipment arrived on the scene, fronted by Chrysalis producer Sean Doherty. They claimed to be part of a small Marlborough- or Swindon-based company which was in the area by chance and started interviewing the observers. The saucer had been visible for 4-5 minutes before passing over the village. While the APRA members were quite cool about the event, some of the witnesses reacted in a fairly predictable manner, expressing surprise. Interestingly, none were able to gauge the size of the UFO accurately, with one observer suggesting it had a diameter of 100ft (30.5m).

The revised flight path had taken the saucer towards the west and it was no longer possible to say with absolute certainty where it would come down. Nigel Blake and an assistant were positioned along the Winterbourne Monkton road and now gave chase in their van. Having flown for a distance of about 3 miles (4.8km), the model landed in a field near Avebury Trusloe and was hastily recovered by Blake and his colleague in near darkness.

With the stunt completed, Chrysalis had plenty of useful video footage in the bag, but they were unhappy with the response by APRA members. This resulted in a production meeting on 23 September when it was decided to remove APRA's presence at the 'Red Lion'. It is also understood that the idea of undertaking a more elaborate UFO hoax in America was briefly discussed, but nothing came of this. The TV programme was fairly

Above: **The Chrysalis flying saucer undergoes a series of secret test flights in a large hangar at RAF Aston Down. Note the spoiler added to the craft's underside.** *Nigel Blake*

Left: **This map of the Avebury area shows the launch and landing sites for the Chrysalis flying saucer.** *Bill Rose*

Left: **An illustration of the Chrysalis flying saucer as it skimmed over the Wiltshire countryside in 2003, during the staging of** *A Very British UFO Hoax*. Bill Rose

moorings in the backyard and their young son Falcon Heene (who could not be found) was probably in the lower utility section.

The 20ft (6.1m) diameter balloon was crudely fabricated from plastic sheeting covered in aluminium foil held in place with duct tape, and it was fairly apparent from the outset that the helium was unlikely to provide sufficient lift to take off with a small passenger.

Having conveniently videotaped the lift-off, Richard Heene then called the FAA, the police and the KUSA-TV station at Denver, claiming that a runaway balloon might be carrying his son. The authorities took the incident very seriously, with local police in pursuit, eventually assisted by National Guard helicopters. Having drifted across several counties and caused the shutdown of Denver International Airport, the balloon finally came to rest in a field 60 miles (96km) away. Its utility compartment was found to be empty.

Richard Heene was already well known as a publicity-seeking storm chaser and a UFO enthusiast with a belief that mankind is descended from an alien race. The family is also known to wider American TV audiences, having appeared several times on the TV reality show, *Wife Swap*. Once the balloon had been recovered and the boy was discovered to have been hiding in the rafters above the garage, suspicions were raised about the authenticity of the incident. Nevertheless, the story attracted international interest and the Heenes were interviewed by Wolf Blitzer on CNN's *Larry King Live*. Falcon Heene was asked by Blitzer, "Why did you not come out of the garage?" He then turned to his father and said, "You guys said that, um, we did this for the show."

On 18 October, the Sheriff's Office announced that they were convinced this incident had been staged and the Heenes would almost certainly face a number of charges. The official response to the incident had been time-consuming and expensive, with *The New York Times* suggesting it had cost approximately $2 million. On 13 November 2009, Richard Heene pleaded guilty to the charges brought against him and was sentenced to ninety days in custody and one hundred hours of community service. Mayumi Heene also admitted her guilt and received a twenty-day weekend jail sentence. There was also a hefty fine imposed and restrictions on profiting from the story for four years.

well received and as a piece of film-making it worked well. The production was visually interesting and occasionally quite amusing.

However, there were a number of obvious problems with the hoax, which failed to fool the APRA members. The change in flight path, instability of the craft, noisy motors and degree of directional control spoilt an otherwise well-executed project. Nevertheless, bearing in mind the time allowed to design, build and test the model, this was a very impressive effort by the Cutting Edge team.

During the original research for this story, Nigel Blake told the author that he believed the original design could be significantly improved and the problems with control, endurance and motor noise could largely be eliminated.

THE HEENE STUNT

Another UFO hoax involving a helium-filled balloon took place on 14 October 2009 at Fort Collins, Colorado, USA. Unfortunately, this publicity stunt was ill conceived, badly executed and involved a six-year-old boy who was feared to have been carried aloft in a runaway balloon. The organisers of this affair were Richard and Mayumi Heene, who claimed that their homemade flying saucer shaped balloon had escaped from its

Index